MW00901447

Marine Corps Reference Publication

MCRP 3-10A.4

Marine Rifle Squad

August 2020

U.S. Marine Corps

United States Government

DISTRIBUTION STATEMENT A: Approved for public release; distribution is unlimited.

PCN 144 000290 00

UNITED STATES MARINE CORPS

7 August 2020

FOREWORD

Marine Corps Reference Publication (MCRP) 3-10A.4, *Marine Rifle Squad,* contains tactics, techniques, and procedures for rifle squad employment and captures lessons learned from recent decades of conflict. It covers a range of missions that Marine rifle squads deal with today or may deal with tomorrow. Despite this extended coverage, MCRP 3-10A.4 is not intended to be all-encompassing. There are many aspects of rifle squad operations that are not detailed herein because a specific publication already exists for that purpose and covering it would be redundant. It also does not address force structure and weapon system changes anticipated over the next several years unless sufficient experimentation and testing have already been conducted to generate best practices that are ready to be codified here. This publication does serve, however, as the basic warfighting squad publication and is to be used in conjunction with the appropriate infantry platoon, company, and battalion publications, as well as other publications that relate to the conduct of ground operations.

This publication is intended for Marine Corps infantry company and below leadership, including units serving as provisional infantry. It is a foundational document that assists in the preparation and execution of squad-level operations in the current operational environment in order to accomplish their assigned missions.

This publication cancels MCIP 3-10A.4i, *Marine Rifle Squad,* dated 10 June 2019 with change 1 dated 15 May 2020.

Reviewed and approved this date.

SCOTT A. GEHRIS
Colonel, U.S. Marine Corps
Commanding Officer, Marine Corps Tactics and Operations Group

Publication Control Number: 144 000290 00

Marine Rifle Squad

Table Of Contents

Chapter 1. The Nature of the Rifle Squad

Chapter 2. Planning

Chapter 3. Offensive Operations

Chapter 4. Defensive Operations

Chapter 5. Patrolling

Chapter 6. Military Operations on Urbanized Terrain

Chapter 7. Movement

Appendices

Glossary

References

CHAPTER 1
THE NATURE OF THE RIFLE SQUAD

MISSION AND PURPOSE

The mission of the rifle squad is to locate, close with, and destroy the enemy by fire and maneuver, or repel the enemy's assault by fire and close combat.

The Marine Corps rifle squad is the fundamental maneuver unit of the Marine Corps infantry. It is organized to provide multiple fire units and mutually supporting combined arms effects on the battlefield. Time and again, the tide of a battle has been changed through a squad leader's decisive employment of their squad. This publication's purpose is to serve as the primary guide for squad leaders to organize, train, and employ their squads. This publication is not meant to be an all-inclusive document regarding the finer details of the rifle squad's employment across the range of Marine air-ground task force (MAGTF) missions. Where necessary, this publication points to other publications which may supplement the material contained herein.

WARFIGHTING PHILOSOPHY

Maneuver warfare is the fundamental warfighting doctrine of the Marine Corps. Marine Corps Doctrinal Publication (MCDP) 1, *Warfighting*, defines maneuver warfare as—

> "...a warfighting philosophy that seeks to shatter the enemy's cohesion through a variety of rapid, focused, and unexpected actions which create a turbulent and rapidly deteriorating situation with which the enemy cannot cope."

The tables of organization and equipment for the Marine rifle squad are specifically designed to achieve the effects of maneuver warfare at even the tactical small unit level. Mission tactics is a fundamental enabler of maneuver warfare. Mission tactics is a method of command which relays commander's intent to subordinate leaders and empowers them to take initiative on a dynamic battlefield in order to achieve that intent in a manner best fitting the realities facing the subordinate leader. Rifle squad leaders are the point at which mission tactics are actually employed in the operational environment. They must be able to effectively and efficiently use the tools and techniques available to them in order to generate the desired commander's intent. To that end, the Marine Corps has developed small unit tactics and techniques which have proven effective for small unit leaders in past wars and operations to serve as a template to guide squad leaders as they face new and challenging environments. The content of this publication should be learned and mastered by squad leaders in order to ensure they are basically able to succeed in scenarios they are likely to confront while leading Marines in any clime and place. However, a foundational aspect of Marine Corps leadership is the small unit leader's ability to understand the fundamentals of the profession of arms and confidently apply them in innovative ways. This ability is the core enabler of mission tactics and maneuver warfare.

ORGANIZATION

The rifle squad totals 13 Marines and is led by a squad leader. Its Marines are divided into three similarly outfitted fire teams of four Marines each, led by fire team leaders (refer to figure 1-1).

Squad Leader

The squad leader carries out the orders issued by the platoon commander. This Marine is responsible for the discipline, appearance, training, control, conduct, and welfare of the squad at all times, as well as for the condition, care, and economical use of weapons and equipment. In combat, squad leaders are responsible for the tactical employment, fire discipline, fire control, and maneuver of their squads. They take position where they can best carry out the orders of the platoon commander and observe and control their squads.

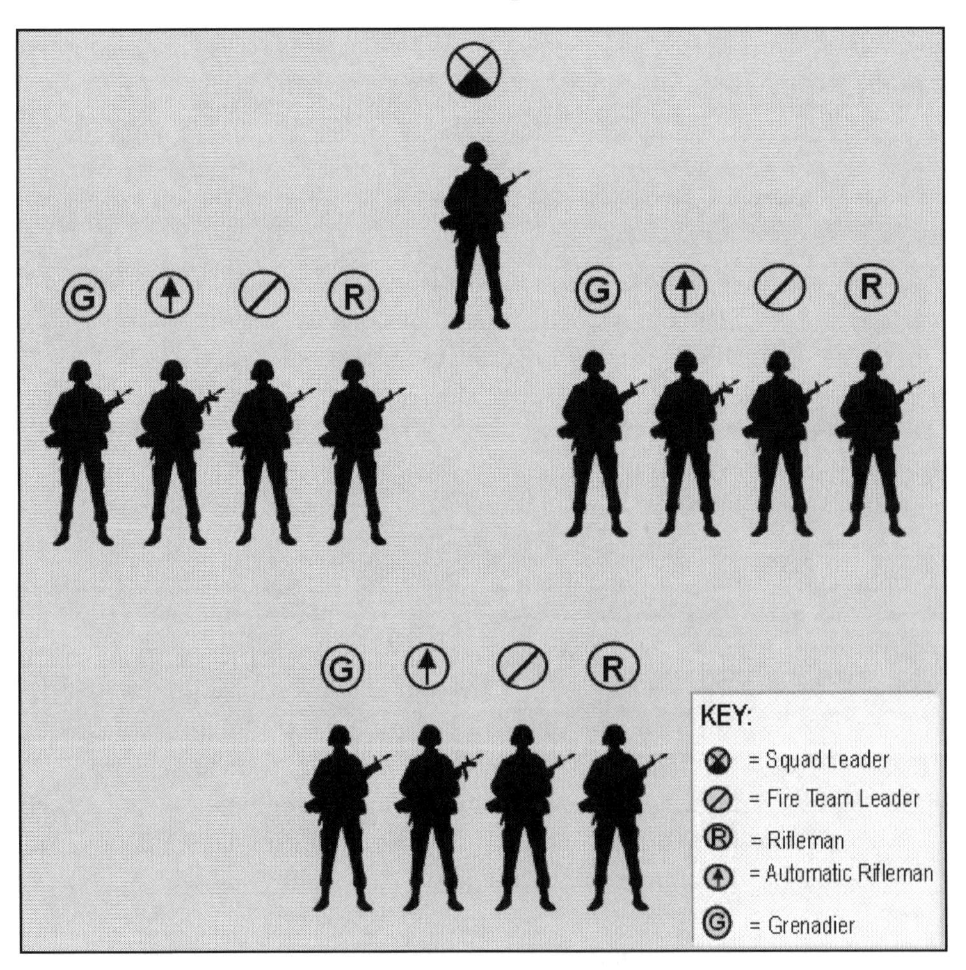

Figure 1-1. Marine Rifle Squad.

Fire Team Leaders

Fire team leaders carry out the squad leader's orders. They are responsible for the fire discipline and control of their fire teams, and for the condition, care, and economical use of their weapons and equipment. In carrying out the orders of the squad leader, they take up the position that will best enable them to observe and control their fire teams. They are typically close enough to their automatic riflemen to exercise effective control of their fires. The senior fire team leader in the squad serves as assistant squad leader. In this role, this Marine sees to the administrative management of the squad in order to allow the squad leader to focus on the squad's tactical employment. Should the squad leader become incapacitated or unavailable, the assistant squad leader assumes the squad leader's roles and responsibilities.

Grenadiers

The grenadier marksmen carry out the orders of their fire team leader. They are responsible for the care of their weapons and equipment. Each grenadier's armament is supplemented with a grenade launcher. They are responsible for the effective and economical employment of their weapons to create battlefield effects that support the squad leader's scheme of maneuver and target precedence.

Automatic Riflemen

The automatic riflemen carry out the orders of their fire team leader. These riflemen are responsible for the effective employment of the automatic rifles and for the condition and care of their weapons and their equipment. Their responsibility during combat is to provide accurate and appropriately prioritized suppressing and assault fires to support the squad leader's scheme of maneuver and target precedence.

Riflemen

The riflemen carry out the orders of their fire team leader. They are responsible for the effective employment of their rifles and for the condition and care of their weapons and equipment. Their responsibility is to deliver accurate rifle fire on the enemy in accordance with the squad leader's scheme of maneuver and target precedence.

Designated Marksman

With the appropriate training and equipment, a senior squad member may be assigned as the squad's designated marksman. The designated marksman carries out the orders of the squad leader and is responsible for the effective employment (to include care and cleaning) of the precision rifle and equipment. The designated marksman provides additional situational awareness to the squad through the use of advanced optics and observation techniques and can also provide accurate rifle fire at longer ranges than associated with riflemen, according to the squad leader's scheme of maneuver and target precedence. The designated marksman's location in the squad's formation is determined by the squad leader, based on the plan that would best match the overall scheme of maneuver.

Preparation for Combat

Effective preparation for combat is an essential task for the squad leader. The squad's success or failure is often dependent on the squad leader's strong leadership and attention to detail. Squad leaders should (at a minimum) conduct both pre-combat checks and pre-combat inspections prior to departing friendly lines. Upon returning to friendly lines, the squad leader's top priority should be returning the squad to its maximum combat readiness by refitting and conducting additional pre-combat inspections, regardless of whether or not additional mission orders have been received. Appendix A contains example checklists for pre-combat checks and pre-combat inspections. All inspections should be extremely detail oriented; a squad leader should be uncompromising in their standards at all times. Making exceptions to unit standing operating procedures (SOPs) or relaxing the standards for equipment and personnel readiness may put the mission and the lives of the Marines in the squad at unnecessary risk.

APPLICATION OF FIRE

The organic weapons within the rifle squad are lethal by their very nature. However, when the weapons of the Marine rifle squad are employed in a coordinated and disciplined manner, their overall battlefield effects increase exponentially. It is critical for squad leaders to train and employ their units with the utmost attention to the disciplined and controlled employment of their assigned weapons. In combat, it is the job of the squad leader to employ the whole of the squad's lethal potential to force the enemy into a dilemma. If employed properly, the Marine rifle squad is capable of offering the enemy only courses of action that will lead to their defeat.

Fire Commands

Since enemy personnel, like Marines, are trained in the use of cover and concealment, threats are often indistinct or invisible, seen only for a short time, and rarely remain uncovered for long. When a threat is discovered, leaders and squad members must define its location rapidly and clearly. When the squad or fire team leader has made a decision to fire on a threat, they direct and control the fires of their fire units by issuing fire commands. The elements of the fire command are *alert, direction, distance, range, assignment, and control*. The acronym ADDRAC is often used to refer to the fire command itself. While leaders will determine the method of engagement and controls, it is also incumbent on members of the squad to issue the core elements of fire commands in order to aid the squad's situational awareness. It is impossible to plan which Marine will spot the threat first.

The following are the six elements that comprise the fire command and a description of each:

- *Alert*. The fire unit is alerted to be ready to receive further information. The alert may also tell "who" is to fire.
- *Direction*. The general direction of the threat is given. When possible, the compass points or azimuth should be provided for direction.
- *Description*. The intended threat is described.
- *Range*. Range information is given to set the sight or adjust the point of aim.
- *Assignment*. This identifies who is to fire on the threat and at what rate of fire.
- *Control*. Commanders signal to open fire.

Fire command examples:

A *Squad*
D *Direct front*
D *Machine gun emplacement*
R 300 *meters*
A *Second team, rapid rate*
C *At my command*

If time does not allow for a full ADDRAC, squad members should provide at minimum the direction, description, and range of enemy targets. Squad leaders should train their Marines to take the initiative in developing the squad's situational awareness during an engagement.

Subsequent Fire Commands

Subsequent fire commands are used by the squad leader to change an element of the initial fire command or to cease fire. To change an element of the initial fire command, the squad leader gives the alert and then announces the element to change. Normally, the elements that require changing are the assignment or control.

Geometries of Fire

When engaging the enemy, the squad must take steps to maximize the geometry of their fires with respect to the target (see figure 1-2). The squad's fires are classified as one of the following types:

- *Frontal fire*. The long axis of the beaten zone is at a right angle to the long axis of the threat.
- *Flanking fire*. Fire delivered against the flank of a threat.
- *Oblique fire*. The long axis of the beaten zone is at an angle to the long axis of the threat (but not a right angle).
- *Enfilade fire*. The long axis of the beaten zone coincides (or nearly coincides) with the long axis of the threat. This class of fire is either frontal or flanking, and is the most desirable class of fire with respect to the threat because it makes maximum use of the beaten zone, and therefore offers the highest probability of the squad's fires impacting their targets.

Characteristics of Fire

The squad leaders' knowledge of application of fires must include the anticipated action and effect of the projectiles when fired. This section discusses various characteristics of direct and indirect fire weapons.

Cone of Fire. Each bullet fired from a rifle at the same threat follows a slightly different path, or trajectory, through the air. The small differences in trajectory are caused by slight variations in aiming, holding, trigger squeeze, powder charge, wind, or atmosphere. As the bullets leave the muzzle, their trajectories form a cone-shaped pattern known as the cone of fire (see figure 1-3).

Beaten Zone. The cone of fire striking a horizontal threat forms a beaten zone which is long and narrow in shape. Beaten zones on horizontal threats vary in length. As range increases, the length of the beaten zone decreases. The slope of the ground affects the size and shape of the beaten zone. Rising ground shortens the beaten zone; ground sloping downward at an angle less than the curve of the trajectory lengthens the beaten zone. Ground that falls at an angle greater than the fall of the bullets is in defilade and will not be hit.

Minimum Safe Distance. Every weapon in the US inventory has been extensively tested, along with their ammunition variants, to determine the necessary safety factors associated with their use. It is imperative that squad leaders are mindful of and train their Marines on the associated minimum safety factors associated with the employment of their assigned weapon systems.

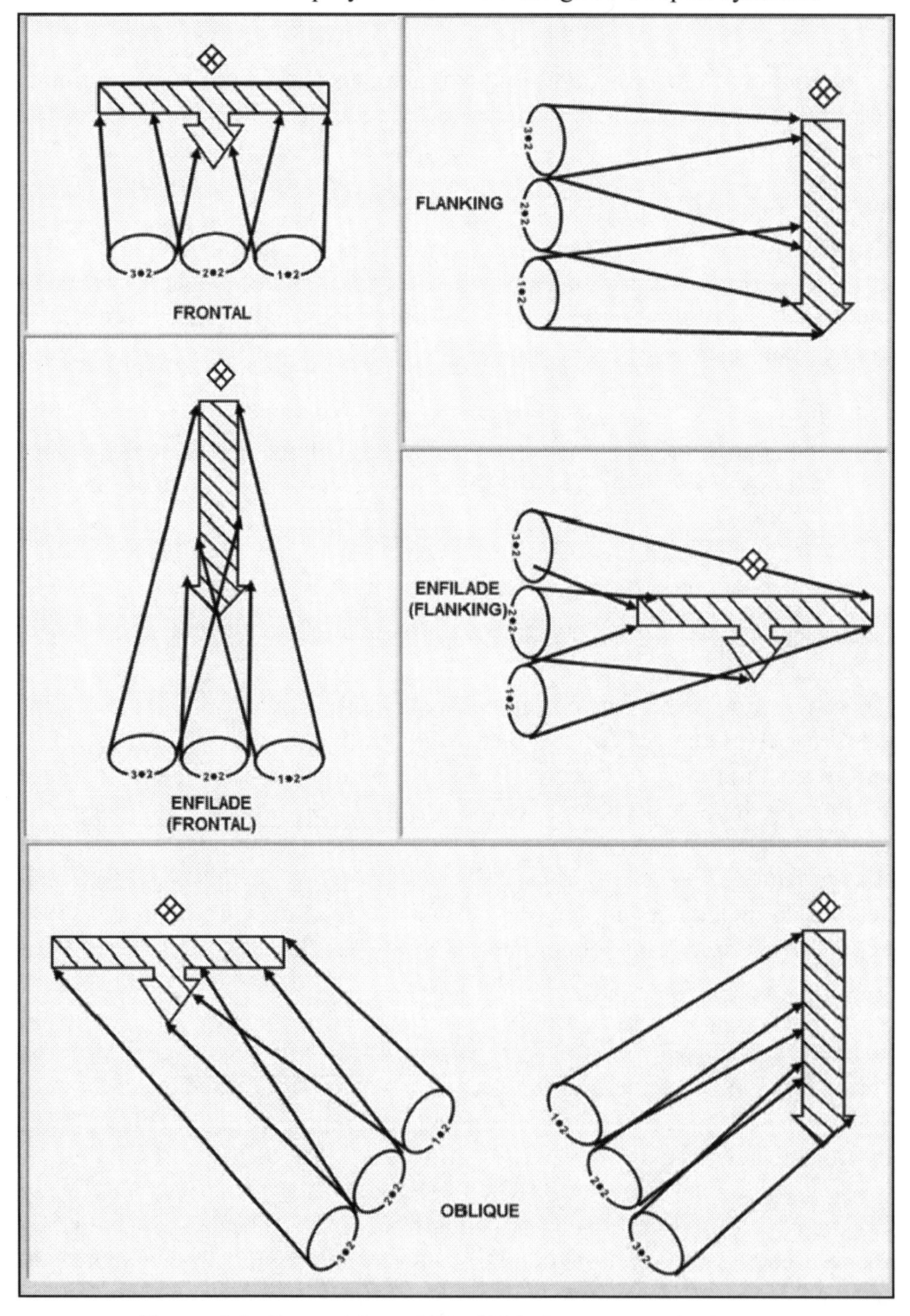

Figure 1-2. Geometries of Fire With Respect to the Target.

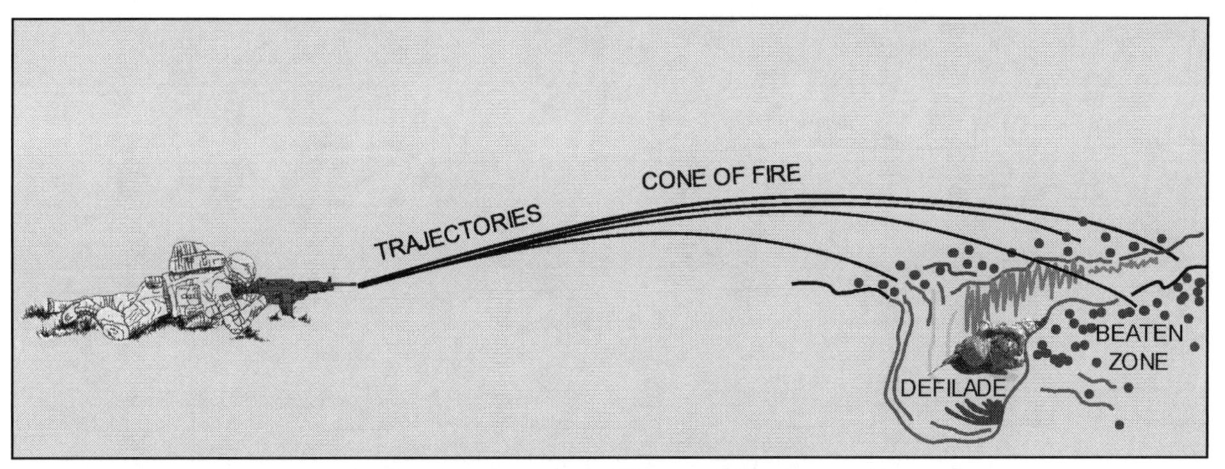

Figure 1-3. Cone of Fire and Beaten Zone.

Squad leaders should be aware of the effects of explosive munitions, friendly or otherwise, on their squads as they maneuver. Closing to *danger close* distances will alter the manner in which the squad calls for fire and may impact the larger scheme of maneuver. Additionally, closing with the enemy while receiving supporting fires requires the squad to recognize how far they can advance before calling on those fires to cease or shift. Failure to pay attention to the effects of fires may negatively impact the friendly scheme of maneuver and put Marines in danger. Additional information on these values may be found in Marine Corps Reference Publication (MCRP) 3-31.6, *Multi-Service Tactics, Techniques, and Procedures for the Joint Application of Firepower (JFIRE)*.

Being within the *danger close* range of munitions does not necessarily restrict the use of those munitions. However, additional procedures will be taken by the firing agency to ensure the accurate delivery of fires when employed close to maneuvering troops. The *minimum safe* distance represents the closest a Marine in full body armor in the prone position may be to the point of impact without incurring unacceptable risk of incapacitation (see figure 1-4). It should be noted that even within the minimum safe distance of a munition, mitigating terrain may sometimes be utilized to move even closer to the intended target.

Warning

Extreme care and close coordination with firing agencies are critical to mitigating the considerable risks of closing with the impacts of munitions.

The minimum safe distance friendly personnel should be from the gun-target line (GTL) is dictated by three factors: the squad leader's risk assessment, the distance from the maneuvering unit to the GTL, and the angle between the position of the maneuvering unit and the GTL. Range safety for the Marine Corps can be found in Marine Corps Order 3570.1C, *Range Safety.*

Warning

The minimum safe distance for live-fire is different in training and combat. Squad leaders should reference the Marine Corps orders that govern range safety for specific geographical locations to gain an understanding of these differences.

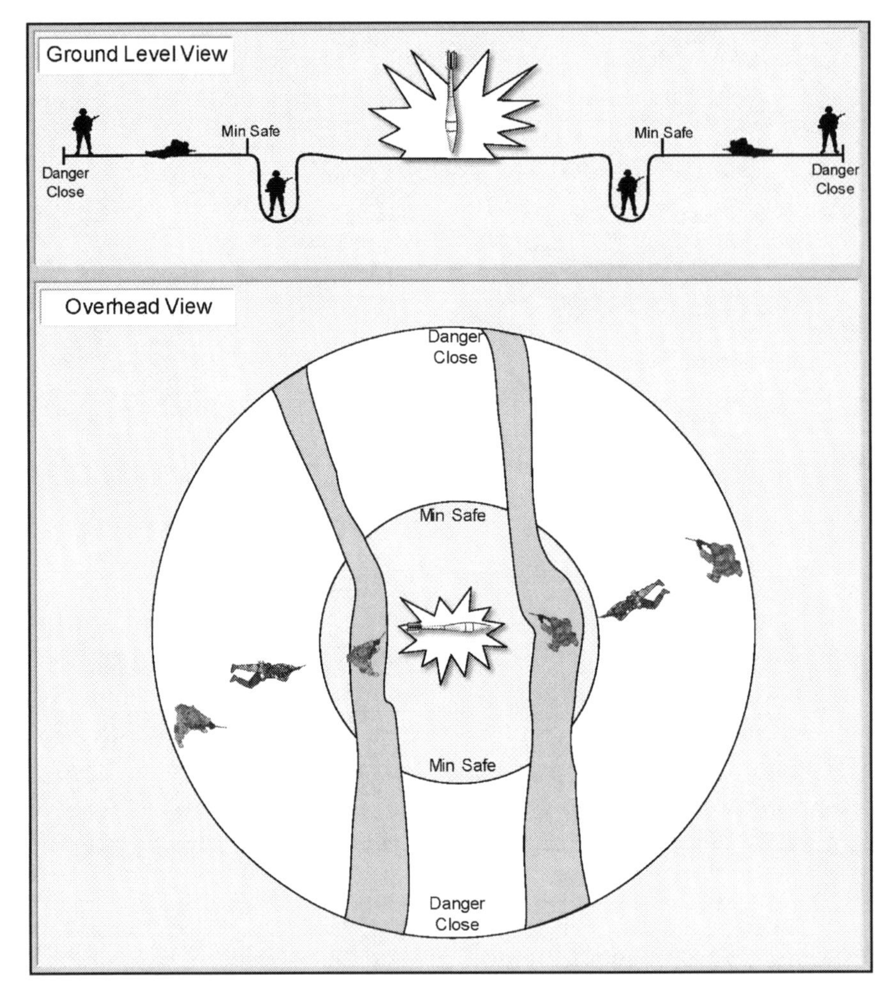

Figure 1-4. Example Minimum Safe Distance.

Squad leaders are responsible for assessing the risk of closing on a target with support from direct and indirect fires. Factors that impact their risk assessment include—

- The nature of the supporting fires and ammunition selected, from direct to indirect to aviation-delivered fires.
- The proficiency of supporting units in delivering and generating accurate fires and effects.
- The effectiveness of supporting unit fires (i.e., effective enemy return fire indicates the need to hold maneuver units until supporting fires are refined).
- Whether terrain will favorably mitigate the effects of friendly fires (e.g., intervening higher ground or maneuver units in defilade).

It is critical that supporting units establish and communicate well-understood right and left lateral limits (see figure 1-5). The implicit trust between supporting and maneuver units requires that any deviations from planned geometries be communicated prior to their implementation so that the maneuver unit is not put in danger.

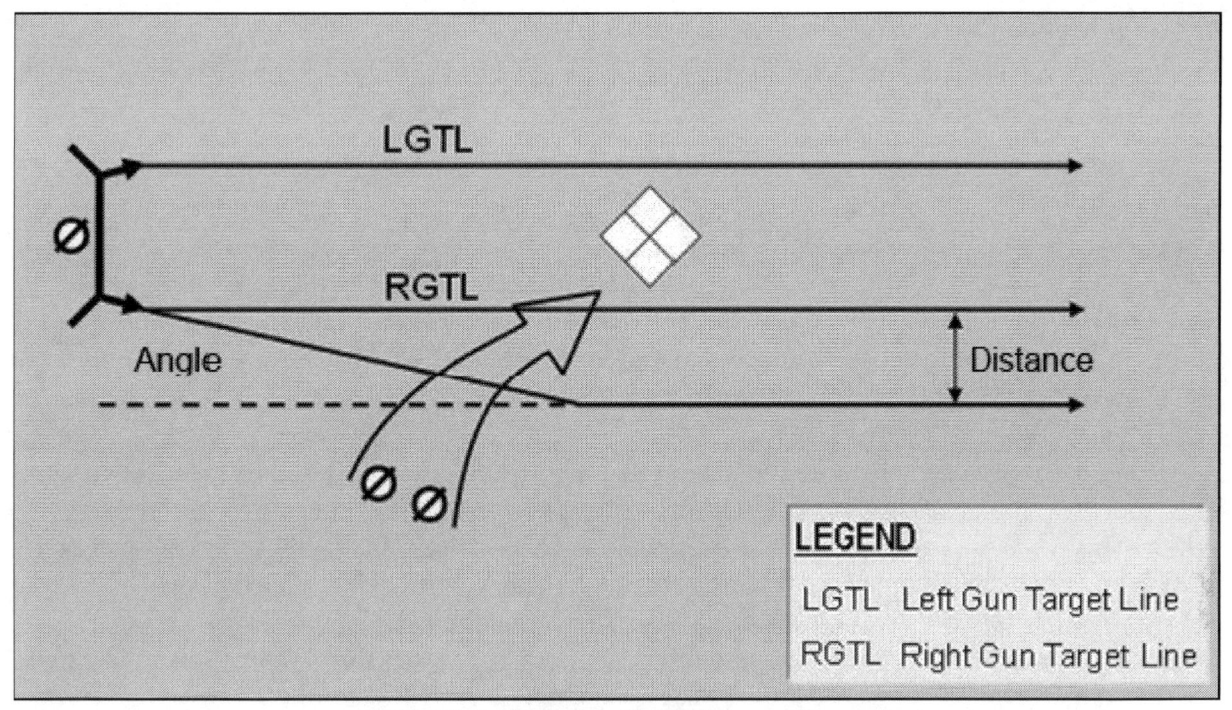

Figure 1-5. Minimum Safe Distance Example.

Rates of Fire and Fire Discipline. The two rates of fire a squad engages with are the sustained and rapid rates. The sustained rate is a rate at which the shooter maximizes the economy of rounds while still accurately and consistently engaging the target and generating desired effects on the target. The more aggressive rapid rate is employed when the unit must overwhelm the enemy by volume of accurate fires and seize or regain fire superiority. However, the rates of fire are ultimately determined by how fast a Marine can accurately engage the enemy and create the battlefield effects consistent with the scheme of maneuver.

Fire discipline is critical to the effective employment of a squad's weapon systems. In combat, the inherent complications of effectively engaging a target are magnified due to human factors such as adrenaline and stress. It is imperative that squad leaders train their Marines to collect themselves during engagements and employ their weapons with the maximum amount of accuracy possible at their assigned rates of fire. The slower, calmly executed, accurate rifle shot is more successful than the panicked burst. Additionally, squad leaders must be mindful that engagements are not necessarily fleeting events. Squad leaders must ensure their squads' resources, including their ammunition, last the duration of the engagement and that they are still prepared for the next one. Maintaining fire discipline is essential to maintaining a unit's combat effectiveness throughout a battle. To this end, squad leaders must rely heavily on their fire team leaders to ensure their Marines are focusing on the fundamentals of accurate weapon employment while following their squad's scheme of maneuver.

Effects of Fires

In addition to simple target destruction, a squad's combined fires can achieve critical effects on the enemy. These effects, which are described in the following subparagraphs, often drive the squad leader's selection of rates of fire in support of the scheme of maneuver.

Fire Superiority. This effect is gained by subjecting the enemy to fire of such accuracy and volume that the enemy fire ceases or becomes ineffective.

Suppression. The goal of suppression is to fix the enemy in position and prevent them from interfering with friendly actions. Accurate and properly distributed fires at a sustainable rate are key to achieving this effect.

Assault Fires. During an assault, the final approach to the objective is often the most dangerous movement for the squad. It is critical that fire superiority is maintained during this time. To achieve this, the squad employs assault fires wherein Marines systematically and aggressively engage any potential enemy position as they make their final approach to the objective.

Combined Arms Effect. Because the squad has both direct and indirect fire weapons available to it, it is possible to achieve combined arms effects where both direct and indirect fire weapon systems are able to affect the target simultaneously. The resulting dilemma has a devastating effect on the enemy and may be critical to their defeat. An example of this combined arms effect would be if the automatic riflemen in a squad were able to fix an enemy unit in place in an entrenched fighting position while the grenadiers fire grenade rounds over enemy cover and into their entrenchments. Then, the enemy is faced with the choice to either stay in place and die or attempt to run and be engaged by the squad's riflemen delivering accurate shots at individual targets. Neither option is particularly attractive, and the dilemma may cause the enemy to break.

Unit Fires

An integral part of the application of fires is the ability to control and apply the squad's combined fires. The squad must effectively apply their fires as a unit to eliminate threats. The application of unit fires provides the squad and/or team leader with the ability to apply fires to both area and point threats. A fire unit is a group of Marines whose combined fire is under the direct and effective control of a leader. The fire units discussed in this manual are the Marine rifle squad and its fire teams. The size and nature of a threat may call for the firepower of the entire fire unit or only parts of it. The type of threat suggests the type of unit fire to be employed against it.

Distributed Fire. Distributed fire is fire spread in width and/or depth to keep all parts of the threat under fire (see figure 1-6). Each Marine fires their first shot on the portion of their threat/sector that corresponds to their position in the squad. They then distribute their remaining shots over the remainder of the threat, covering that portion of the threat on which they can deliver accurate fire without changing position.

Concentrated Fire. Concentrated fire is fire delivered from a unit at a single threat (see figure 1-7). A large volume of fire is delivered at the threat from different directions, causing the beaten zones of the fire unit's weapons to meet and overlap and giving maximum coverage on the threat. For example, an enemy automatic weapon that has gained fire superiority over an element of a particular unit can often be neutralized by concentrated fire from the remaining elements which are not under direct fire.

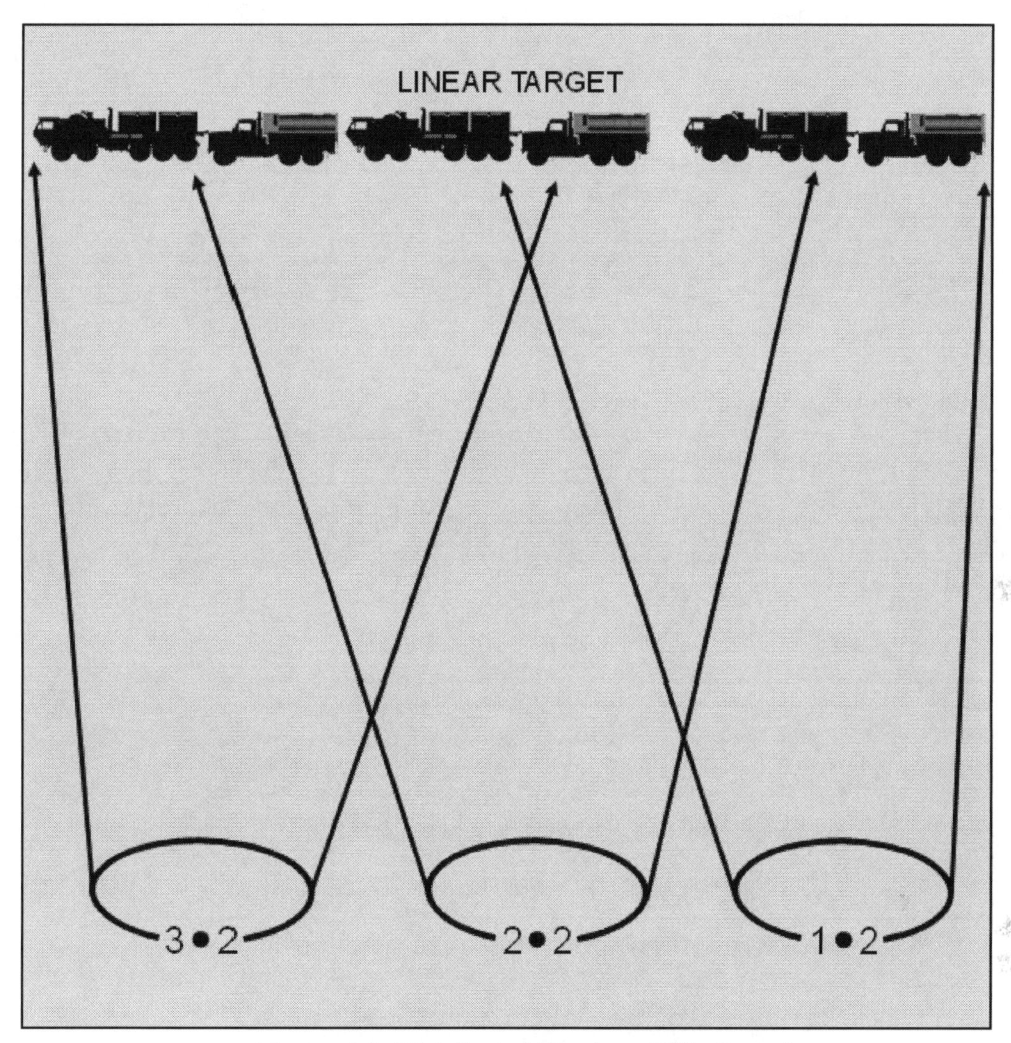

Figure 1-6. Distributed Fire by a Rifle Squad.

Combinations of Concentrated and Distributed Fire. The fire team organization of the Marine rifle squad permits the squad leader to combine both concentrated and distributed fire in engaging two or more threats at the same time (see figure 1-8). As an example, the leader of a squad delivering distributed fire on a threat could shift the fire of one fire team to engage an alternate threat with distributed fire on a smaller target area.

Natural or man-made terrain features, fractions (i.e., dividing the threat area in half or thirds), and target reference points can all be used to ensure fires are distributed across the width and/or depth of a point or area threat.

Weapons-to-Target Match

The organic weapons of the rifle squad enable it to achieve combined arms effects on the enemy and provide the tools to address specific threats on the battlefield. Weapon-to-target matches will ultimately be established by the orders of the squad leader; generally, the target priorities depicted in table 1-1 are the most common.

Table 1: 1. Squad Organic Weapons-to-Target Match.

Weapon System	Target Precedence
Rifle/Carbine	Point targets and enemy infantry
Grenade Launcher	Crew-served weapons, targets in defilade, clusters of enemy troops, troops in the open
Automatic Rifle	Crew-served weapons, clusters of enemy troops, point targets
Light Antiarmor Weapons	Light armored vehicles, crew-served weapon emplacements, fortified positions
Fragmentation Grenades	Crew-served weapon emplacements, fortified positions, clusters of enemy troops

Range Determination

Range determination is a process of determining the approximate distance from an observer to a target or other distant object. Accurate range determination allows the squad members to set their sights correctly and place effective fire on enemy targets. Three methods of determining ranges are estimation by eye, use of available optics, and observation of fire.

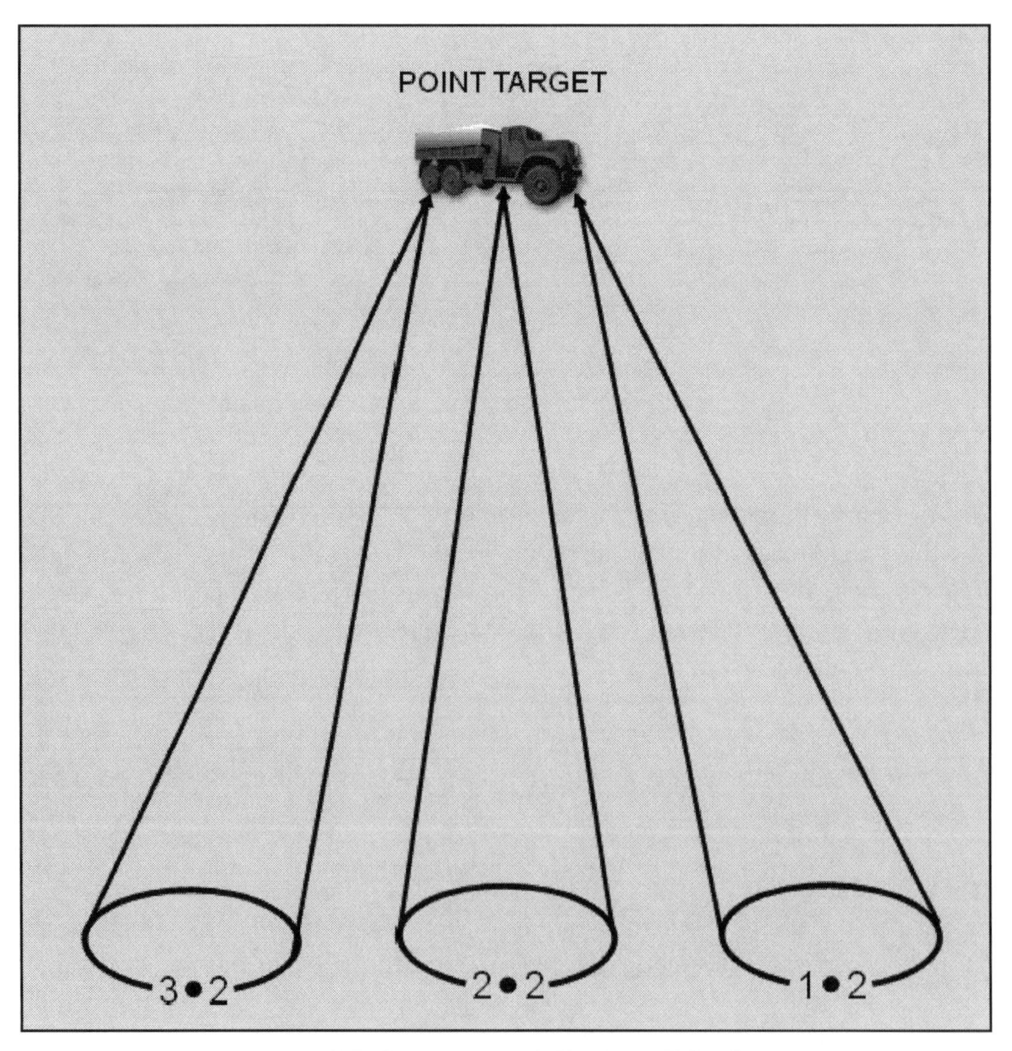

Figure 1-7. Concentrated Fire by a Rifle Squad.

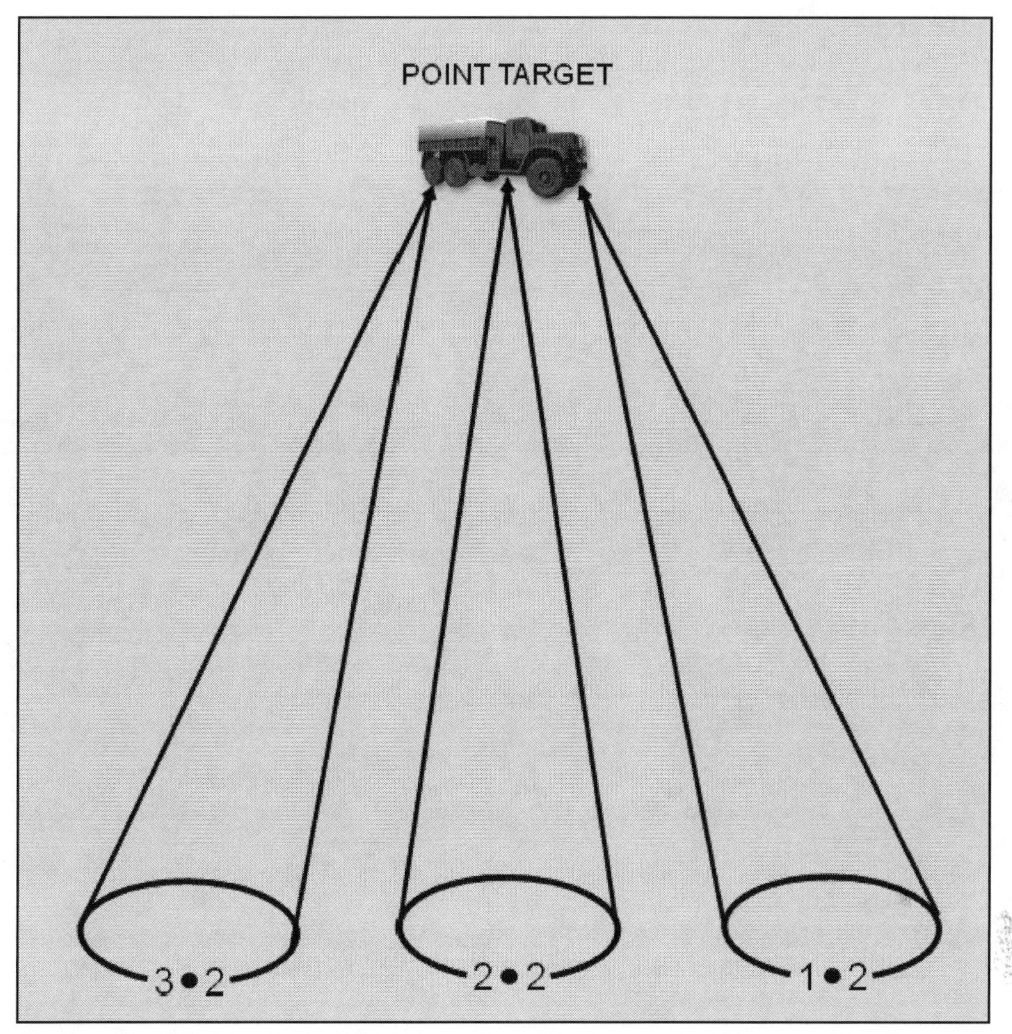

Figure 1-8. Combination of Distributed and Concentrated Fires on Two Targets.

Estimation by Eye. There are two methods used in estimating range by eye—the mental unit of measure and the appearance of objects. With training and practice, accurate ranges can be determined and a high volume of surprise fire can be delivered on the enemy.

Mental Unit of Measure. To use the mental unit of measure method, visualize a 100-meter distance (or any other familiar unit of measure). With this unit in mind, mentally determine the number of these units between their position and the target (see figure 1-9). In training, mental estimates should be checked by pacing off the distance. The average person takes about 130 steps per 100 meters. Distances beyond 500 meters can most accurately be estimated by selecting a halfway point, estimating the range to this halfway point, then doubling it (see figure 1-10).

Appearance of Objects. When overlooking hills, woods, or other obstacles, or when most of the ground is hidden from view, it is impractical to apply the mental unit of measure method to determine range. In such cases, through practice, Marines can learn to estimate distance by familiarization at various known ranges. For example, watch a person standing 100 meters away. Fix their size and details of their features/equipment firmly in mind. Watch them in the kneeling, and then in the prone position. By comparing the appearance of a person at 100, 200, 300, and 500 meters, one can establish a series of mental pictures. When time and conditions permit, accuracy can be improved by averaging estimates from different individuals to find a range.

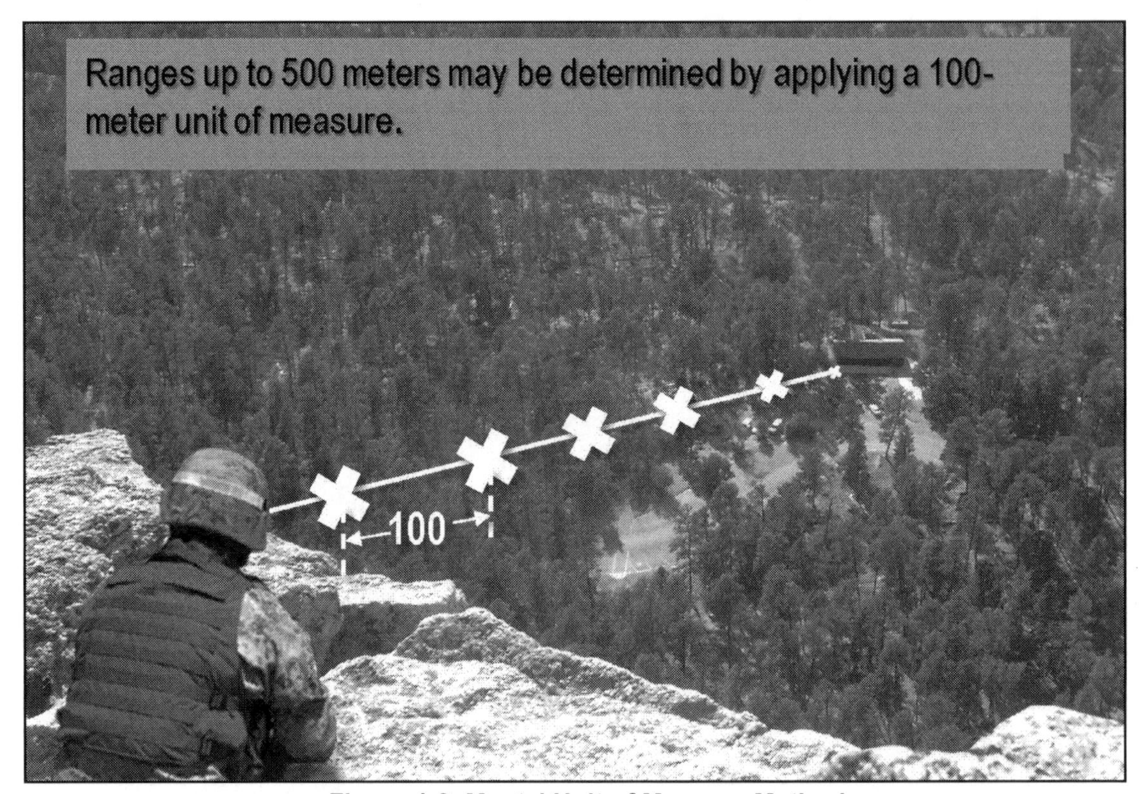

Figure 1-9. Mental Unit of Measure Method.

Figure 1-10. Mental Unit of Measure Method (beyond 500 meters).

Use of Available Optics. A squad's organic optics can assist in determining the range of enemy threats. Sighting systems and laser range finders are typically organic to squads and can be utilized for range determination. To better understand how organic squad optics can assist in range determination, refer to the associated optics technical manual.

Observation of Fire. Accurate range determination can be made by having an observer (not the rifleman) observe the strike of tracer or ball ammunition. In this method, the range may be estimated quickly and accurately; however, the possibility of achieving surprise is lost and the firing position may be revealed to the enemy. This procedure requires that—

- The Marine firing estimates the range by eye, adjusts the point of aim for that range, and fires.
- The observer follows the path of the tracer and notes the impact of the round.
- The observer calls out an adjusted aiming point (or hold), adjusting for elevation and windage corrections to hit the threat.
- The Marine firing adjusts the aim point (or hold), continues to fire, and makes corrections until a round hits the target. The Marine relays the final aim point used and announces "hold."

Line Company Weapons Platoon Capabilities

A squad might be augmented by other units that are organic to the infantry company. Squad leaders must familiarize themselves with the capabilities and limitations of these units (refer to table 1-2), and, when appropriate, incorporate their leadership's expertise into the creation of the squad's scheme of maneuver. This cooperative planning is even more important for partnering with attachments from the battalion or other units assigned to the squad.

Setting Conditions

The goal of the fire application techniques discussed above is to generate a desired battlefield effect in order to set the conditions for decisive tactical action. During planning and rehearsals, Marines should be able to identify specific battlefield conditions which—once achieved—implicitly drive follow-on actions.

For example, as the squad is in the assault, the assaulting fire teams should be able to recognize lulls in enemy incoming fire from effective fire suppression. This would drive the fire team leaders to transition to assault fires, rapidly close the remaining distance to the objective, and engage in close combat. If, however, they realize effective suppression has not been achieved, the fire team leaders should take the initiative to increase their rates of fire or improve their battlespace geometries to set the required conditions for the next tactical action. In this example, the fire team leaders are empowered and acting as fighter leaders, and the squad leader has maintained responsive, decentralized control.

Table 2: 2. Weapons Platoon Capabilities and Limitations.

Unit	Capabilities	Limitations
Medium Machinegun Squad	Provides a high volume of fire for a sustained period. May engage targets at greater ranges than the squad due to their weapon systems and stabilizing tripods. Machine guns generally come with more advanced optics than are available to the squad and provide increased situational awareness.	The additional ammunition and heavier weapon systems bring additional weight which limit the unit's maneuverability and speed. Machine guns are best employed in pairs and should not be split up in the same manner the squad would divide fire teams.
Mortar Squad	The mortar has the longest range of the organic weapons in the rifle company. The higher angle of fire allows the mortar squad to engage targets without exposing friendly forces to the enemy's direct fires. Different ammunition types such as smoke and illumination rounds allow greater flexibility and provide unique options to the scheme of maneuver.	The mortar weapon system is heavy and awkward to carry and may restrict the unit's mobility. There is a premium on the amount of ammunition that can be carried due to its size and weight. Each round must be used wisely.
Assault Squad	Assault squads provide the best option for neutralizing or destroying armored or entrenched targets at close to medium range. The demolition expertise of infantry assault Marines can be used in explosively breaching obstacles or the destruction of heavy equipment.	Size and weight limits the number of available rockets. Contingencies must exist for scenarios where rocket supply is insufficient to neutralize or destroy their targets. Rocket-associated back-blast may impact the squad's scheme of maneuver.

Movement and Maneuver

The organization of the rifle squad and the manner in which it closes with its objective are important aspects of the squad leader's scheme of maneuver. The squad leader will determine which fire team and squad-level formations are best suited for the situation. Additionally, the squad leader deploys the squad in the assault using fire and movement and fire and maneuver to maximize the effects of the unit's fires within the constraints of all relevant geometries of fire.

Base Unit

A fundamental method of controlling the movement and maneuver of the rifle squad is the employment of the base unit concept. During mission planning or as part of unit SOP, the squad leader designates the base unit. The base unit is almost always one of the fire teams. While all fire team leaders are responsible to know and execute the squad leader's scheme of maneuver, by using a base unit, the squad leader can execute implicit control of the squad during movement and maneuver. The base fire team is used by the squad leader to control the squad's direction, position, and rate of movement. The other fire teams do not need to maintain rigid positions in relation to the base fire team; the base fire team is used as a general guide. If another fire team can move forward more rapidly than the base fire team, it should do so. For instance, if the base fire team is receiving enemy fire, but the terrain in front of another fire team provides cover from enemy fire, the latter team should move rapidly forward to a position where they can deliver fire on the enemy. Covering the base fire team's movement by fire takes pressure off them and permits them to move forward. Once the base fire team comes generally abreast, the other fire teams can then resume fire and movement. This allows the squad leader to focus on the overall mission rather than the sum of the smaller tactical actions taking place in the squad, while also allowing fire team leaders to take the initiative by controlling their teams in a manner best suited to the squad leader's intent and the changing nature of the battlefield.

Fire Team Formations

Each fire team leader typically determines the formation for their own units. Thus, a squad may contain a variety of fire team formations at any one time, and these formations may change frequently. The relative position of the fire teams within the squad formation should be such that one will not mask the fire of the others. It is not important that exact distances and intervals be maintained between fire teams and individuals as long as control is not lost. Sight or voice contact should be maintained within the fire team and between fire team leaders and the squad leader. All movement for changes of formation is usually by the shortest practical route. The characteristics of fire team formations are similar to those of corresponding squad formations. The characteristics of the fire team formations are as follows.

Fire Team Column. Figure 1-11 displays the fire team column. The characteristics of this formation include the following:

- Permits rapid, controlled movement.
- Favors fire and maneuver to the flanks.
- Vulnerable to fire from the front and provides the least amount of fire to the front.

Fire Team Wedge. Figure 1-12 displays the fire team wedge. The characteristics of this formation include the following:

- Permits good control.
- Provides all-round security.
- Formation is flexible.
- Fire is adequate in all directions.

Fire Team Skirmishers. Figure 1-13 displays the fire team skirmishers formation. The characteristics of this formation include the following:

- Provides the greatest volume of fire to the front.
- Used when location and strength of the enemy are known, during the assault and the pursuit.

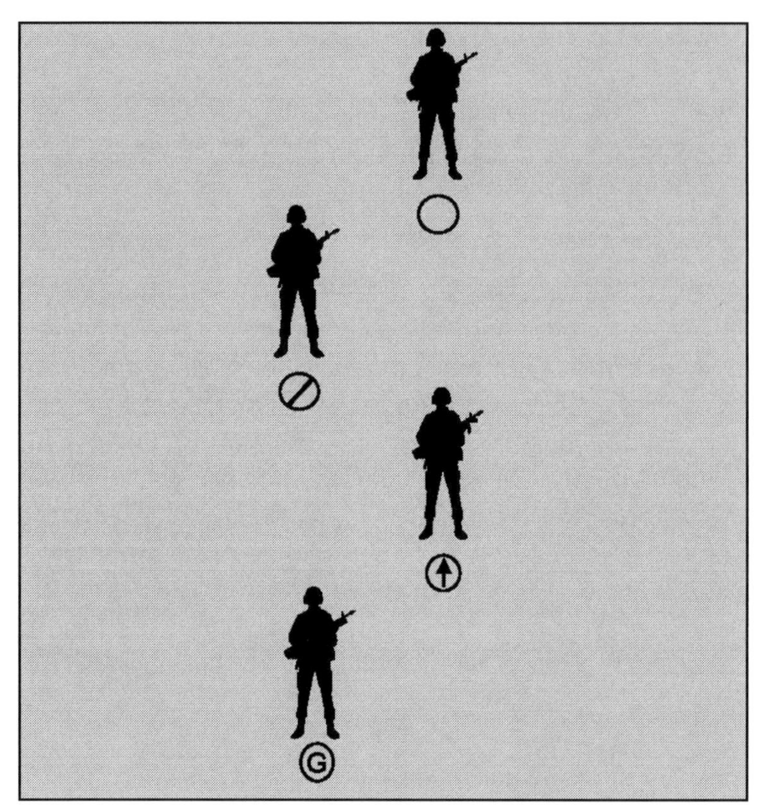

Figure 1-11. Fire Team Column.

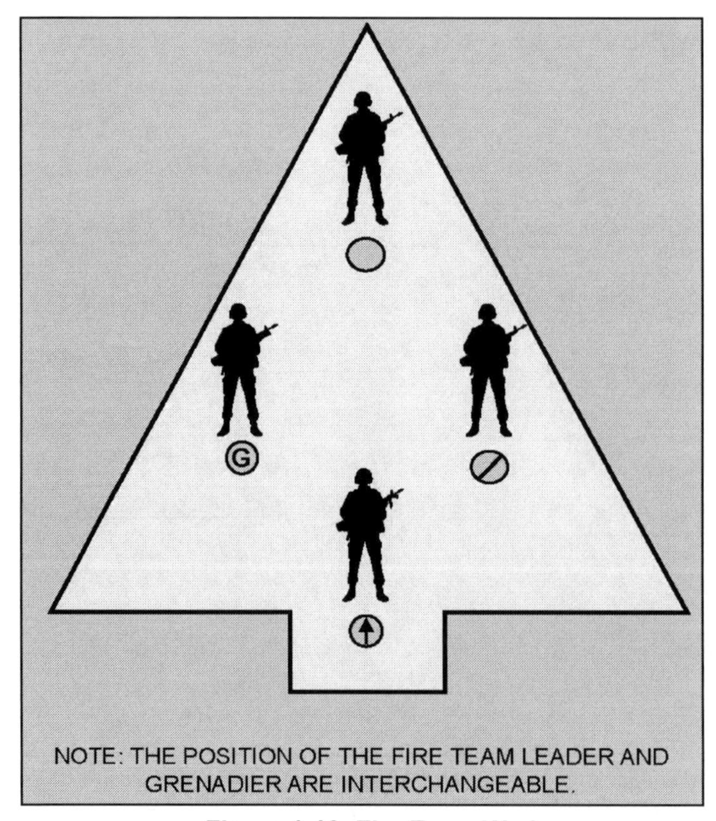

Figure 1-12. Fire Team Wedge.

Figure 1-13. Fire Team Skirmishers (left and right).

Fire Team Echelon. Figure 1-14 displays the fire team echelon formation. The characteristics of this formation include the following:

- Provides firepower to the front and the echeloned flank.
- Used to protect the flanks of a unit.

Squad Formations

The squad leader prescribes the formation for the squad. The platoon commander and squad leader may prescribe the initial formation for their respective subordinate units when the situation dictates or the commander so desires. The subordinate unit leaders may make subsequent changes as the mission progresses and the situation changes. The characteristics of squad formations are similar to those of the fire team.

Squad Column. Figure 1-15 displays the squad column. The characteristics of this formation include the following:

- Easy to control and maneuver.
- Excellent for speed of movement or when strict control is desired.
- Used for narrow covered routes of advance, maneuvering through gaps between areas receiving hostile artillery fire, moving through areas of limited observation, and moving under conditions of reduced visibility.

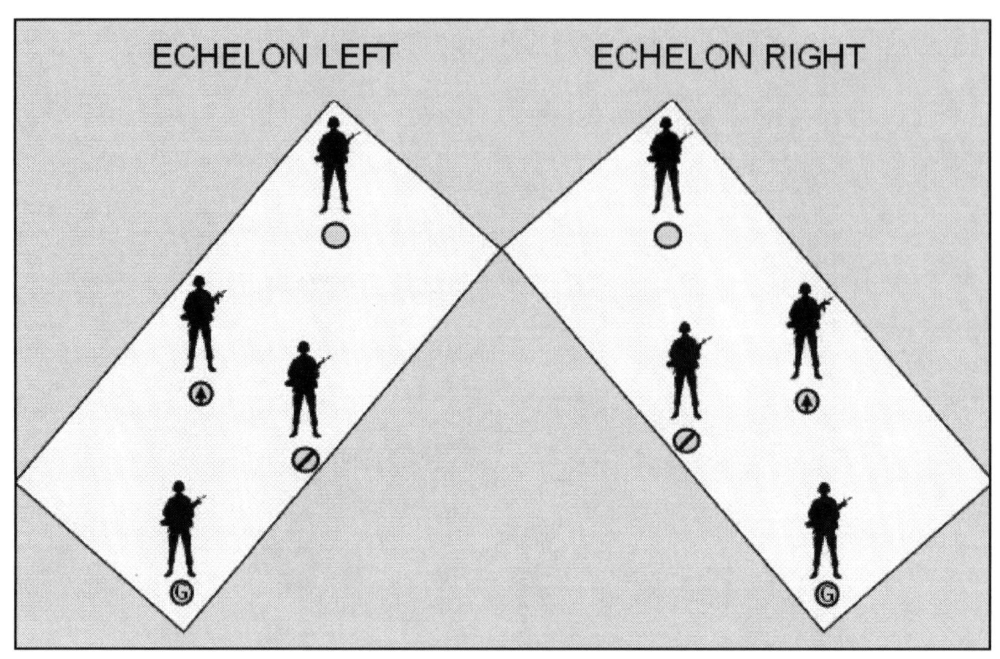

Figure 1-14. Fire Team Echelons (left and right).

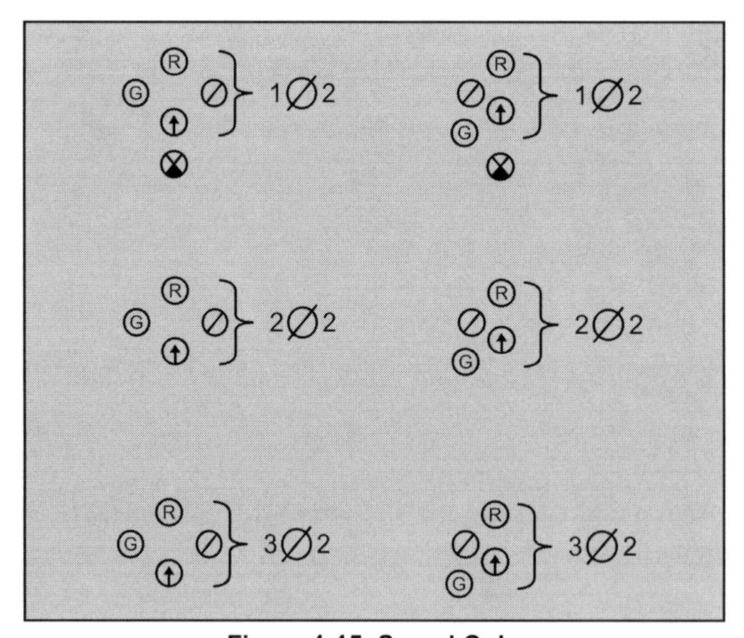

Figure 1-15. Squad Column.

Squad Wedge. Figure 1-16 displays the squad wedge. The characteristics of this formation include the following:

- Improved control/ease of communication.
- Provides all-round security.
- Formation is flexible.
- Fire is adequate in all directions.

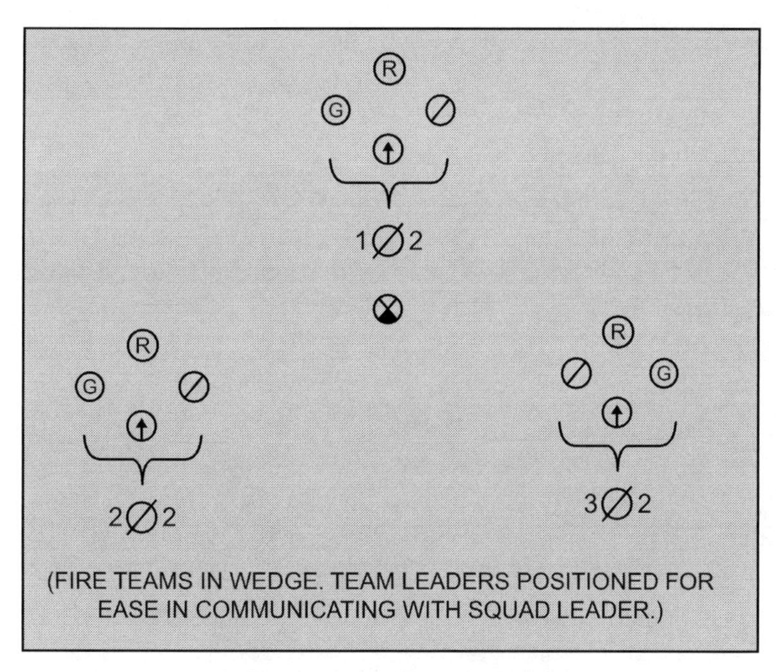

Figure 1-16. Squad Wedge.

Squad Vee. Figure 1-17 displays the squad vee. The characteristics of this formation include the following:

- Facilitates movement into squad line.
- Flexible formation in the assault, which keeps a fire team ready to maneuver on contact from the more protected rear of the formation.

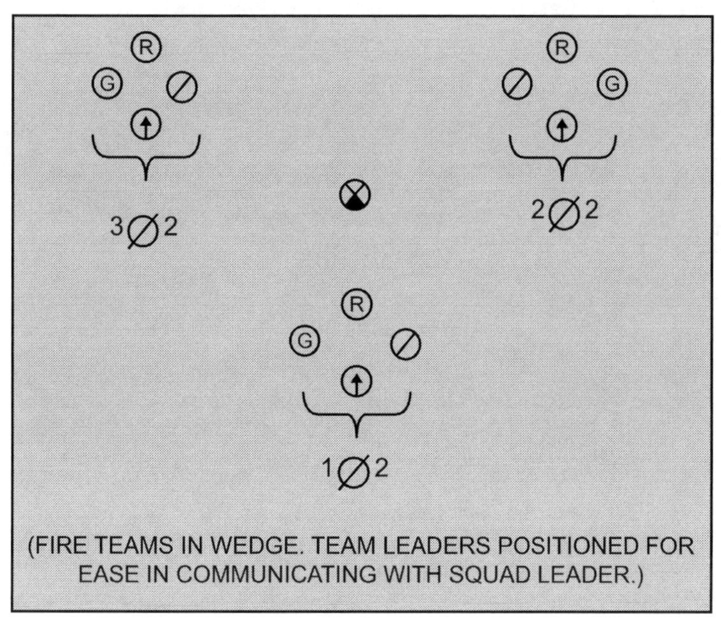

Figure 1-17. Squad Vee.

Squad Line. Figure 1-18 displays the squad line. The characteristics of this formation include the following:

- Provides maximum firepower to the front.
- Used when location and strength of the enemy are known, during the assault and the pursuit.

Squad Echelon. Figure 1-19 and figure 1-20 display the squad echelon. The characteristics of this formation include the following:

- Provides firepower to the front and the echeloned flank.
- Used to protect the flanks of a unit.

Changing Formations

The squad leader may change formations to reduce casualties from hostile fire, present a less vulnerable target, or get over difficult or exposed terrain. Formation changes in varying or rough terrain are frequent as the squad moves over or around both man-made and natural obstacles, such as rivers, swamps, jungles, woods, and sharp ridges. Fire team leaders should utilize hand-and-arm signals to order a change in formation to ensure complete understanding. The squad leader signals the squad formation to the fire team leaders. Fire teams may be in any fire team formation within the squad formation. Hand-and-arm signals can be found in appendix B.

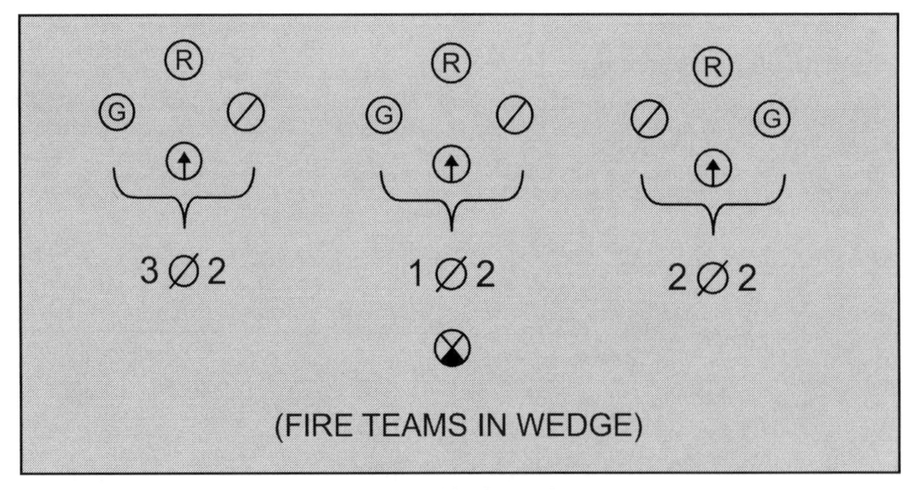

Figure 1-18. Squad Line.

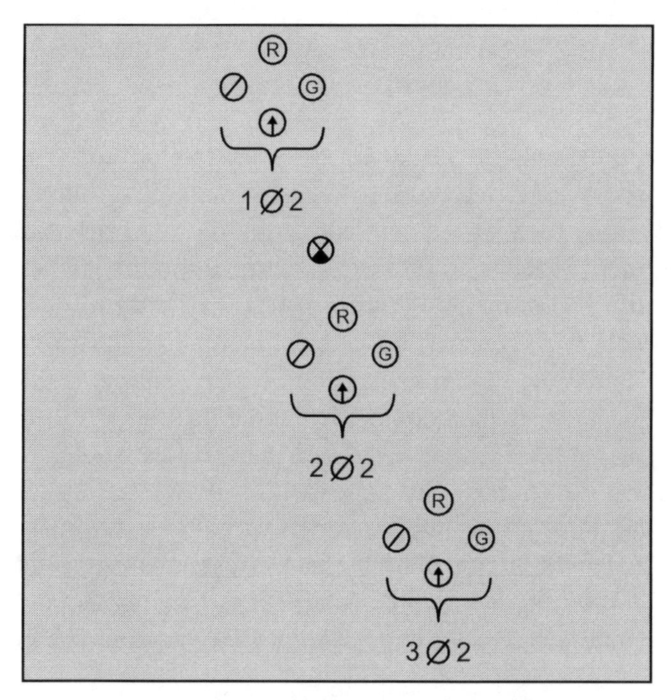

Figure 1-19. Squad Echelon (right).

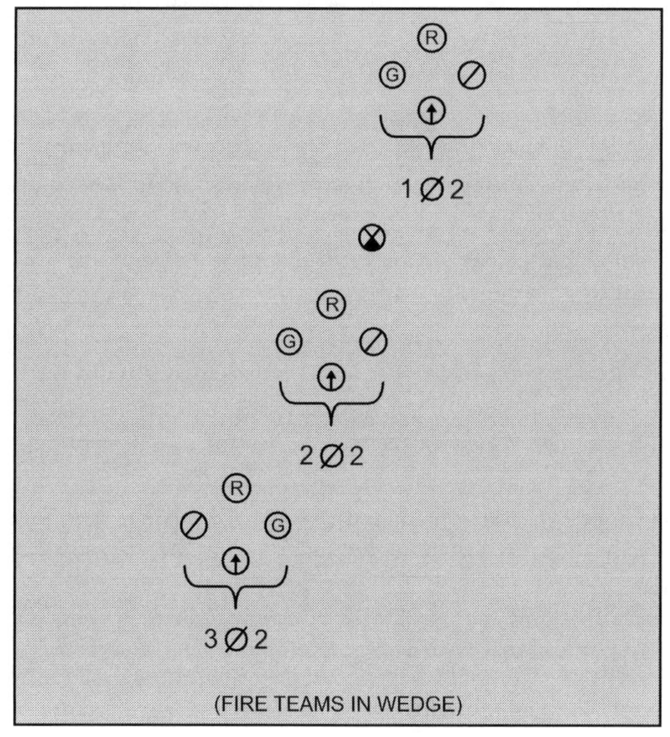

Figure 1-20. Squad Echelon (left).

ACTIONS ON CONTACT

Once contact is made with the enemy, the squad leader will determine whether or not the enemy meets the engagement criteria. If it does, and the enemy's composition and strength is consistent with the information given by the platoon commander's order, the rifle squad must close the distance to the enemy and drive them off or destroy them. To do this, the squad utilizes techniques of fire and movement and fire and maneuver. It is critical that once initiated, contact with the enemy is maintained. To allow the enemy to break contact is to risk them maneuvering on the exposed assaulting unit or allowing them to escape to occupy another position of advantage. The enemy can be fixed and destroyed when contact is maintained.

If the enemy does not meet the engagement criteria, it is crucial that the squad take action to preserve its combat power until it encounters an enemy that does, because the enemy may deploy fire units for the purpose of conducting spoiling attacks, which cause friendly forces to deploy and expend combat power prior to the intended engagement. Squad leaders must temper their units' aggression with tactical patience in order to prevent the enemy from seizing the initiative in situations like this.

Fire and Movement

Once contact is made and the assaulting rifle squad has a known direction to the enemy positions, the squad may begin closing with the enemy directly using fire and movement. This technique, while simple in concept, is difficult in practice. It requires implicit communication between the members of the squad in order to maintain fire superiority to close with and destroy the enemy.

Fire and movement begins with the squad leader deploying the squad into a line formation and issuing an ADDRAC. The squad line formation is preferred because it opens up geometries of fire for each weapon system in the squad to engage the enemy. Depending on the situation, the squad leader may choose to forego maneuver and assault directly into the enemy instead. If this is the case, the squad leader will make it known in the control portion of the ADDRAC.

The cardinal rule of fire and movement is that there can be no movement without fire and there should never be fire without movement. That is to say, suppressing fires on the enemy are required for Marines to expose themselves and rush forward; however, sustaining suppressing fires without taking advantage of them with movement is a waste of ammunition. Because of the volume of fire required to maintain fire superiority in the assault, it is critical that the squad leader compel the squad to move aggressively toward the enemy in order to ensure it is not short on ammunition when it is most exposed.

The most basic yet most crucial element of fire and movement is the proper execution of buddy team rushes. This involves two Marines providing mutual suppression and employing implicit and explicit communication in order to close with the enemy while under fire.

Buddy team rushes begin on contact with the enemy. Both Marines seek cover utilizing micro-terrain while also returning fire in order to gain fire superiority over the enemy. On signal from leadership, the buddy team begins their movement cycle towards the enemy. The rushing Marine assesses that suppression has been achieved, locates the next closest piece of micro-terrain to seek cover behind while maintaining an effective field of fire, and takes a short sprint to it. This sprint should generally take between five to seven steps, minimizing the time the Marine is exposed to enemy fire. Once in position, that Marine assumes the fore position and provides suppression. Once the Marine in the aft position observes that the rushing Marine has achieved effective suppression on the enemy, they begin their rush to the next piece of micro-terrain, ideally beyond the position of the Marine in the fore position. Once the rushing Marine is set behind cover and is able to provide effective suppression, the cycle repeats.

> ***Note:*** It is the responsibility of the Marine in the aft position to be able to swiftly assume the fore position after rushing. All reloading or other adjustments should take place in the aft position to ensure that once in the fore position, the Marine is able to provide continuous suppression on the enemy to protect their fellow Marine as they begin their rush.

After buddy rushes, individual rushes are made that take into consideration the next covered position, the current effects of any covering fire, and Marines' potential exposure to enemy fire. They should also look to put themselves in a position of advantage for possible close combat. This method of closure continues until the unit as a whole is within hand grenade range of the enemy and transitions to assault fires to engage in close combat.

While closing with the enemy, there can be a natural tendency for the squad members to converge on a single point on the enemy's position. As the squad approaches the objective, the converging geometries of fire will cause many of the squad member's sectors of fire to collapse. Squad leaders must be mindful of this and ensure their fire team leaders maintain dispersion and fire team sectors as the squad assaults through an objective.

Fire and Maneuver

The squad leader should look for opportunities to conduct fire and maneuver on the enemy. In contrast to fire and movement, the purpose of fire and maneuver is to establish a base of suppressing fire with one unit while seeking an alternate position of advantage from which to fire and move with another unit. During maneuver, the squad leader must pay particular attention to geometries of fire. The ballistic and explosive characteristics of the squad's weapon systems guide how close the squad leader can bring the maneuvering unit to the suppressing unit's friendly fires.

It is important to note that even the most skillfully executed maneuvers are likely to end in fire and movement and close combat. Maneuver by itself does not guarantee victory. Rather, it allows the squad leader to gain an advantage over a fixed unit and take the initiative by forcing the enemy to react to their decisions rather than reacting to the defending enemy.

Fighter Leader

The success or failure of the Marine rifle squad ultimately depends on the will and leadership of the squad leader. Squad leaders must always seek to position themselves where they can best positively control their units, and at points of friction where direct intervention on their parts may be needed. Squad leaders must keep calm, calculated composure in combat and be ready to lead aggressively from the front, into the face of danger, in order to inspire their Marines and lead by example. While many of the skills of combat leadership can be quantified and drilled, this moral courage must be developed through daily devotion to the Marine Corps and the mission. Leaders at all levels, especially the squad level, must recognize the need to inspire and drive action through their own actions and example. Through this, Marines gain confidence in their leader and the plan, allowing them to overcome the friction around them. Marines will follow a fighter-leader.

CHAPTER 2
PLANNING

PREPARATION

The outcome of combat is often decided well before the first engagement. Marines can fight and win when confronted with crises and unexpected events. However, when possible, a good leader can shape future success well before the first step outside friendly lines. While it is unlikely that all the elements of a well-crafted plan will fall neatly into place in the fog of war, the planning process remains essential to ensuring a unit is as prepared for its mission and its likely contingencies as it can be.

TROOP LEADING STEPS

The six troop leading steps aide squad leaders in their planning efforts (see figure 2-1). They can be remembered using the acronym, BAMCIS.

Begin Planning

Upon receipt of a task from higher authority, whether it be in the form of a warning order or a full combat order, the squad leader should immediately begin the planning cycle. From the onset, time will be a resource in short supply. No plan will ever perfectly match the mission it is given, and leaders should endeavor to complete their planning to allow the maximum amount of preparation time by their subordinates. A general guide is the "one third, two thirds rule," where a leader takes one third of mission preparation time to develop and issue orders and direct rehearsals, while allowing two thirds of the time to be used by subordinate leaders to conduct their own planning and preparation for combat.

BAMCIS

B – Begin Planing

A – Arrange Reconnaissance

M – Make Reconnaissance

C – Complete Planning

I – Issue Order

S – Supervise

Figure 2-1. Troop Leading Steps.

The most effective way to ensure that all of the following steps in the planning process are budgeted with the appropriate amount of time allocation is to employ reverse planning. To do this, the squad leader sets the timeline to prepare for combat, starting with the time to depart friendly lines. By working backwards and inserting time for all crucial preparations for the mission, a better respect is gained for the time required for each step necessary to prepare for combat. This also helps the squad leader ensure that early steps do not consume time that is needed later for more complex preparations, such as rehearsals.

As soon as possible, the squad leader issues a warning order to subordinates. At minimum, it should include the assigned mission, the squad's task organization, the distribution of mission essential equipment, and a timeline to prepare for combat. This enables fire team leaders to prepare while the squad leader goes about the necessary steps to complete the plan. Even if the elements of the warning order change before the final plan is issued, this allows the squad to take the initiative in their preparations for combat. See appendix C for an example warning order template. Directing the construction of a terrain model for briefing the order later is highly recommended, and should be a key item in the warning order's preparation for combat timeline.

Estimate of the Situation. At the onset, the squad leader should complete a thorough estimate of the situation to inform the rest of the planning cycle. Analyzing the mission, enemy, terrain and weather, troops and support available—time available (METT-T) is an effective way to go about analyzing the environment in which the mission will take place. Refer to figure 2-2.

METT-T
M – Mission
E – Enemy
T – Terrain and Weather
T – Troops and Support Available
T – Time Available

Figure 2-2. Estimate of the Situation.

Mission. The squad's mission is based on the company's and platoon's missions and the commander's intent. It is imperative that the squad leader fully understand the higher headquarters (HHQ) commander's intent to ensure the squad is able to support the battlefield effects required for the success of the larger plan. Every plan from the Marine expeditionary force to the squad is nested together through commander's intent, and leaders at every level are responsible for the shared success of the MAGTF.

Enemy. A thorough understanding of the enemy is crucial to formulating a course of action. Marines should remember that the enemy is a thinking, breathing human being who wants to win just as much as they do. Understanding the enemy and their capabilities and motivations is crucial to destroying them. Squad leaders use the acronym SALUTE (see figure 2-3) to assist them in understanding threat forces' capabilities and limitations. Through reconnaissance, it is crucial to understand the following:

SALUTE
S – Size
A – Activity
L – Location
U – Unit
T – Time
E – Equipment

Figure 2-3. Analyzing the Threat.

- Size. The total estimated strength of the enemy.
- Activity. What is the enemy's current course of action? This will address their level of security and likely reactions to the friendly course of action.
- Location. The enemy's placement of forces on the battlefield can indicate a great deal about its intent and battlespace geometries. This information also informs the friendly fire support plan.

- *Unit*. Knowing the enemy composition and unit type can indicate a great deal about their will and military capability. The course of action used against elements of a mechanized infantry brigade from a peer adversary will vary from one against non-uniformed fighters in a violent extremist organization.
- *Time*. Planning must account for the fact that the environment is constantly changing. The enemy or the civilians in the area of operations do not cease action simply because they cannot be observed. Knowing the elapsed time since the last observation of the objective and the time required to close with it, an understanding of the risk level and changes in assumptions about the objective may be formed.
- *Equipment*. Understanding the enemy's equipment and its intended employment informs the friendly course of action. Knowing the enemy's equipment is not enough, however. Squad leaders must also understand their tactics, techniques, and procedures to predict what reactions the enemy will have to the friendly course of action. The general state of their equipment may indicate the morale and readiness of the enemy, as well.

The most critical aspect of making estimates about the enemy comes through "turning the map around." You must assume the enemy forces are just as motivated to defeat you, as you are to defeat them. With that in mind, leaders should explore what they would do if they were on the other side of the battlefield and wanted to defeat the Marine unit. This simple thought exercise of seeing the battlefield from the enemy's perspective offers crucial insights.

Terrain and Weather. The physical environment can also be as much an adversary to the mission as the enemy itself. It can also be utilized to achieve advantages over the enemy. The selected course of action should be enabled by the environment, not a victim of it. The acronym OKOCA (see figure 2-4), can guide analysis of the environment as follows:

OKOCA

O – Observation

K – Key Terrain

O – Obstacles

C – Cover and Concealment

A – Avenues of Approach

Figure 2-4. Environment Analysis.

- *Observation and fields of fire*. Leaders must understand the factors which may enable or impede observation of the battlefield. This can include ambient weather conditions, topography, and vegetation. Observation should also account for the optics and sensors available to both friendly and enemy units.
- *Key terrain*. Key terrain features are locations on the battlefield that play a direct role in the success of the mission. These can be positions for support by fire positions, natural obstacles impeding the assault plan, critical civilian infrastructure, or others. Key terrain is not identified by its size or relative position on the battlefield, but rather by the effects it may have on the accomplishment of the assigned mission.
- *Obstacles*. Natural or man-made obstacles restrict or deny maneuver within the urban area. Bridges, walls/fences, canals, streams, rivers, and rubble created by the effects of weapons or threat activities should be thoroughly analyzed. Construction sites and commercial operations such as lumber yards, brick yards, steel yards, and railroad maintenance yards are primary sources of obstacle and barrier construction materials.

- *Cover and concealment*. Infantry should develop an innate sense of identifying and differentiating cover and concealment. Cover provides protection from the effects of fire. Concealment only provides protection from enemy observation. Analysis of the battlefield should preemptively identify which locations offer either one, or both, well before the first round is fired.
- *Avenues of approach*. An environmental analysis will likely identify natural lanes through which the unit can move and maneuver. The squad leader should select those that offer the best range of tactical options while also providing maximum cover and concealment. However, it is important to recognize that the enemy will likely come to the same conclusions about the battlefield. Therefore, a squad leader should be careful when selecting the path of least resistance, as that path is the one most likely to be covered by enemy fires.
- *Weather*. While weather may prove to be as unpredictable as the enemy, it is important that leaders account for its effects in mission planning. Assuming the worst possible weather should guide preparations. Additionally, leaders may find that the weather may offer distinct battlefield advantages, such as providing concealment where none previously existed, or degrading the enemy's capability and morale during adverse weather conditions.

Troops and Fire Support Available. Once a squad leader has an understanding of the enemy's capabilities, they must understand what friendly capabilities it will take to defeat them. Accounting for attachments, weapon systems, and the available Marines to complete a mission dictates the tactical employment of a squad. The course of action may also be significantly aided by fire support from adjacent or higher echelon units. Requesting these resources early and integrating their employment into the plan from the onset will increase the likelihood of success in the chosen course of action.

Civil Considerations. Battles do not take place in isolation. They occur in the same operational environment occupied by civilians and noncombatants. Leaders must take the time to understand the effects on the civilian population and noncombatant actions on the friendly course of action. All available effort should be taken to minimize the risks to civilians. The United States fights from the moral high ground. Marines' actions in combat should reflect the values and beliefs that form the core of the United States' identity. Disregard for civilians could not only put the tactical mission at risk, but also the honor and ultimate success of the Nation.

Arrange for Reconnaissance

Conducting reconnaissance requires a planning cycle within a planning cycle. Squad leaders must allocate the appropriate time to ensure their chosen form of reconnaissance is achievable. Patrols, requests for information from the company level intelligence cell (CLIC), unmanned aircraft system (UAS) availability, and other collection assets must be coordinated in advance of their use. Prior to conducting reconnaissance, the squad leader must decide their priorities for information collection. By this point, they should have a rough concept of their course of action. The squad leader should take the time to list the assumptions they made for the success of the chosen course of action. They should then seek to objectively confirm or deny these assumptions through the chosen reconnaissance method.

Make Reconnaissance

Conducting a reconnaissance prior to completing the final planning for a mission ensures that the squad leader's plan is not created in a vacuum separate from the reality of the environment, the enemy disposition, and civil considerations. Many methods of conducting reconnaissance are available; the squad leader will need to choose which combination best suits the timeline and assigned mission. The most common methods include—

- *Physical (leader's) reconnaissance*. The squad leader takes a small detachment to physically observe the objective area. The squad leader typically leaves the senior fire team leader to continue supervising preparations for combat, as well as contingency plans in the event that the leader's reconnaissance become compromised or takes casualties.
- *Map/imagery study*. It may not be practical to conduct a physical leader's reconnaissance due to time, asset availability (e.g., vehicles or aircraft), or risk factors. If resources allow, imagery and other products (e.g., debriefs from previous patrols) provided by the CLIC can provide the squad leader an adequate view of key areas of interest.
- *UAS employment*. Resources permitting, employing UASs with their available sensors can significantly increase a squad leader's situational awareness of their area of operations. Live feeds of the objective area reduce the risk incurred by the time elapsed since the last observation by reconnaissance patrols or intelligence products covering the objective. However, similarly to conducting a leader's reconnaissance, care must be taken not to tip off the enemy to friendly intentions through overt observation. An effective deception plan should be used to ensure that if the enemy observes the UAS, they are not alerted to the true target of observation.

Complete the Plan

After conducting the reconnaissance, the squad leader updates their plan with the new information they received. It is unlikely for every question to be answered at this point, but there is never such a thing as a perfect plan. While the squad continues the preparation for combat timeline established in the warning order, the squad leader completes the combat order. For an example order, see appendix C.

Because combat orders are issued to enable mission tactics, as discussed in chapter 1, it is crucial for the squad leader to clearly outline the HHQ's mission and commander's intent, as well as their own. Armed with a good basic order and commander's intent, the Marines will be equipped with the knowledge required to accomplish the mission when friction and changes occur.

Issue the Order

Squad leaders must ensure that their plans are communicated to their Marines in the most effective and comprehensive way possible. The use of terrain models or imagery is highly recommended to allow Marines to visualize the mission as it is briefed. The order is issued to whomever the squad leader deems necessary, ranging from only the key leaders of key elements to everyone involved in the operation. Simply briefing the order, however, is not enough. The order's key elements must be understood and the commander's intent must be internalized by all members of the unit.

Supervise

If time has been managed properly up to this point, the squad leader has only consumed one third of the unit's total time allotted to prepare for combat. Supervising the final preparations for the mission requires the squad leader to exercise direct leadership as they check on the squad's developing readiness. The final supervision should be the pre-combat inspection by the squad leader. During this, the squad leader should not only inspect equipment, but also inspect the Marines' knowledge of the mission and the roles each Marine will have in it. See appendix A.

COMMAND AND CONTROL PLANNING

Primary, Alternate, Contingency, and Emergency Plans

The ability to effectively command and control the unit while coordinating with higher and adjacent units is a key requirement for maneuver warfare. Leaders should put serious thought into formulating their communications plan by identifying the methods of communication for each part of the mission. These methods can be remembered using the acronym PACE—primary, alternate, contingency, and emergency. These choices should be informed both by force protection elements as well as the effectiveness of the communication method in question. On approach to an objective, a squad leader may select hand-and-arm signals as the primary method with radios as contingency or emergency options in order to ensure the unit is not detected in the electromagnetic spectrum prior to the assault. Likewise, other situations may call for satellite-enabled data links to be the primary communication method, with pyrotechnics as emergency signals for distributed operations. The PACE plan should be developed for every part of the plan, and squad members should memorize it, including changes during the flow of the mission.

Attachments and Detachments

In addition to communications on the battlefield, leaders must plan for command relationships during the course of operations. While the chain of command for doctrinal infantry units is well understood by those within them, supporting units are commonly attached, and organic elements detached, to support the concept of operations. Operational control of all these elements lies with higher echelon headquarters, but tactical leaders must clearly define the roles for tactical control of their units.

Unity of command cannot be in question during combat. For example, if an electronic warfare team is attached to an infantry squad, the squad leader will likely maintain tactical control of them, regardless of the rank or parent organization providing the electronic warfare team. Similarly, if a squad is directed to detach a fire team to provide security for a medium mortar section, the gaining mortar squad will effectively have tactical control over that team, and the squad leader providing the fire team no longer has direct control over it.

Fire Support Planning

Planning for fire support during the course of a mission requires technical rigor and advance planning and coordination. Fire support can be a critical enabler for mission success or it can pose a significant risk of fratricide. Fire support integration is a critical element to the Marine Corps' maneuver warfare philosophy and can allow a numerically inferior force to create battlefield effects well beyond its organic means. At a minimum, squad leaders should understand the following elements:

- *Friendly, civilian, and enemy positions*. The most essential element of fire support planning is knowing the locations of the relevant parties on the battlefield. Due to the high risk of fratricide, it is unacceptable to employ fire support assets without knowing friendly positions. Likewise, employing fire support without reasonable knowledge of the enemy wastes valuable resources. Finally, knowledge of local patterns of life and tracking civilian and noncombatant elements on the battlefield can affect the fire support available to the mission. The tactical success of a fire mission that destroys the enemy but causes unnecessary civilian casualties may lead to strategic failure.
- *Call for fire reporting formats*. Due to the highly technical nature of fire support, the language used to request and control it is highly standardized. Knowing the most current methodologies for one's area of operations and memorizing their elements is critical training and a pre-combat inspection item for all Marines. See appendix D for basic fire support and other reporting formats.
- *Gun-target line*. This is a straight line drawn between the firing position of a fire support asset and the target location. Far from representing a straight shot, a GTL represents the probable trajectory of the projectile being fired. All weapon systems in the US inventory have been tested to ensure their rounds fall within certain consistent geometries. However, it is important to understand the probable trajectory in three-dimensional space of the ordnance being employed. This will affect the scheme of maneuver by limiting airspace allocation and ground maneuver. Certain fire support assets are reliable for ground elements to maneuver under their GTL. Others, such as light mortars, are generally not considered safe to maneuver under.
- *Minimum safe distance*. The minimum safe distances are those determined through testing the probable effects of munitions in the US inventory. These values, expressed in meters from impact, represent the minimum safe distances at which friendly forces can be positioned without cover and reasonably expect that the ordnance in question will not produce casualties in the friendly force.
- *Enemy threat rings*. In considering the fire support plan, leaders must also take into account the capabilities of known fire support systems in the enemy's inventory. These are communicated through threat rings, which are represented on maps by red circles over enemy fire support emplacements with the radius matching the system's effective range. Leaders must be aware of their spatial relationship to enemy weaponeering effects as they maneuver through the battlespace and recognize the risk to friendly maneuver and support assets.
- *Joint terminal attack controllers and joint fires observers*. To integrate fires from multiple sources in the joint environment, a joint terminal attack controller (JTAC) or joint fires observer (JFO) may be organic or attached to the squad. These are Marines who have received special training to control fires. While these Marines remain under the tactical control of the squad leader, the squad leader should take special care to coordinate and cooperatively plan fire support for the scheme of maneuver with these Marines. This allows the squad leader to fight the overall tactical fight without becoming bogged down in the technical and often time-consuming control of fire support.
- *Information operations considerations*. Warfare is fought not just on the physical domain but also in the information domain. Information operations (IO) and its information-related

capabilities are tools that the squad leader can leverage to accomplish the mission. For more information on information related capabilities, see MCWP 3-32, *Marine Air-Ground Task Force Information Operations*. Commanders may direct that specific elements of IO be integrated into the squad's scheme of maneuver.

- *Communication strategy and operations*. This capability (formerly referred to as combat camera and public affairs) is used to communicate powerful visual images from the battlefield and integrate them into a larger strategic narrative. While it may seem cumbersome to carry camera equipment into combat, squad leaders must understand that powerful visual images are the "supporting fires" of the information domain.
- *Advancing the narrative*. Information operations often hinge on "being first with the truth." This means that as we conduct our own IO, the enemy will be attempting to advance theirs as well. Ensuring that friendly force actions and speech are consistent with the strategic narrative is critical to controlling effects in the information domain.
- *Tactical deception*. Deception involves the use of tactics designed to purposely deceive an enemy and mislead them into taking actions that make them more vulnerable or that give the friendly force an advantage. Deception is an advanced form of IO. It can be intended to create either lethal or nonlethal effects. Lethal effects make the threat susceptible to their effects to force them into unfavorable maneuver. Nonlethal effects may be desired to force threats out of hiding, make them move away from contact, discredit their efforts, or deny them popular support. At times, it may be necessary to conduct feints, demonstrations, or other activities to present the enemy with a narrative that leads them to make disastrous assumptions about friendly forces. Intelligently employed tactical deception can have devastating effects on the battlefield and should be tied to every level of planning.

Force Protection

Complementary to selecting a course of action that achieves the assigned mission, leaders must also take into account plans to preserve and protect friendly forces. This is known as force protection. These are methods which can be employed to lessen the impact of adversary actions on friendly units.

Operations Security. Operations security identifies and controls critical information, including indicators of friendly force actions attendant to military operations, and incorporates countermeasures to reduce the risk of an adversary exploiting friendly vulnerabilities. Controlling information prevents the adversary from gaining insight into friendly force actions through various means.

Emission Control. Emission control is the selective and controlled use of electromagnetic, acoustic, or other emitters to optimize command and control capabilities while minimizing (for operations security) detection by enemy sensors, mutual interference among friendly systems, and/or enemy interference with the ability to execute a military deception plan. Leaders must be aware that their signatures may be detectable in the electromagnetic spectrum. Techniques such as brevity of transmissions, using the lowest required power settings, and terrain masking can be used to enhance protection against enemy electronic warfare systems. Proper emission control procedures may deny the enemy early warning, as well as limit their ability to conduct direction finding activities to target friendly force emissions.

Personal Protective Equipment. Personal protective equipment is equipment worn by each Marine to protect them from the hazards of the battlefield. Leaders must weigh the costs and benefits of personal protective equipment postures against the environment and the threat. Close urban combat may require heavier protection, while long-distance jungle patrols require more mobility. These decisions should come from HHQ mission requirements and not from comfort seeking.

Air Defense. The air domain may often turn out to be contested. In order to account for this, leaders should seek to mask the signature of their units while also employing active defense measures if capabilities are available. The squad's first line of defense is concealment, and Marines should understand the capabilities and limitations of enemy sensors and take steps to frustrate those capabilities. If the squad believes it is under aerial observation, then it should take immediate steps to confuse or deceive the enemy sensors through any means necessary.

Chemical, Biological, Radiological, and Nuclear Defense. Higher headquarters evaluates and monitors the nature of the chemical, biological, radiological, and nuclear (CBRN) threat in an area of operations. Tactical leaders must ensure that their units are educated on the threat and trained and equipped to operate under these adverse conditions. Proper planning in this category may ensure the survival of the unit in otherwise cataclysmic conditions. For more information on CBRN defense, see Marine Corps Tactical Publication (MCTP) 10-10E, *MAGTF Nuclear, Biological, and Chemical Defense Operations*.

CHAPTER 3
OFFENSIVE OPERATIONS

PURPOSE OF THE OFFENSE

The offense is the decisive form of warfare. While other operations might cause great damage to or harm an enemy's interests, offensive operations are the means to decisive victory. Offensive operations allow commanders to impose their will on the enemy by shattering the enemy's moral, mental, and physical cohesion.

CHARACTERISTICS OF OFFENSIVE OPERATIONS

Squad leaders should seek to maximize the characteristics of the offense during all offensive activities. They should strive to gain surprise, achieve concentration by massing the lethal and nonlethal effects of fires, generate tempo, and execute plans with audacity.

FUNDAMENTALS OF OFFENSIVE OPERATIONS

While the characteristics of offensive operations are generalizations, the fundamentals are rules that have evolved over time to accomplish assigned missions through application of the principles of war. Table 3-1 lists the offensive fundamentals.

Table 3-1. Fundamentals of Offensive Operations.

Orient on the enemy	Gain and maintain contact
Develop the situation	Concentrate superior firepower at the decisive time and place
Achieve surprise	Exploit know enemy weaknesses
Seize or control key terrain	Advance by fire and maneuver
Gain and maintain the initiative	Maintain momentum
Neutralize the enemy's ability to react	Act quickly
Provide security for the force	Exploit success
Be flexible	Be aggressive

SQUAD IN THE OFFENSE

Offensive action, or maneuver, consists of five steps: preparation, conduct, consolidation and reorganization, exploitation, and pursuit. Each step is subdivided according to the mission and/or unit involved. Squad leaders can apply these steps to any offensive activity tasked to their squads; but because the mission of the squad is to attack, the following planning steps will be viewed through the lens of a squad participating in an attack. In both planning and execution, some steps may be shortened, omitted, or repeated. The steps pertinent to squad leaders are as follows:

- *Preparation*.
  - Movement to the assembly area.
  - Reconnaissance and rehearsals.
  - Movement to the line of departure.
- *Conduct*.
  - Movement forward of the line of departure to the assault position.
  - Advance by fire and maneuver.
  - Arrival at the assault position.
  - Assault and advance through the assigned objective.
  - Consolidation and reorganization.
- *Consolidation and reorganization*.
  - Establish security.
  - Assume hasty defensive positions.
  - Position any supporting elements.
  - Redistribute arms, ammunition, and supplies.
  - Restore internal communications.
  - Prepare for enemy counterattack.
- *Exploitation*.
  - Continue the attack.
  - Prepare to exploit tactical success.
- *Pursuit*.
  - Prevent the enemy from escaping and destroy them.

Preparation

The preparatory step begins on receipt of the warning order. It ends when the lead element crosses the line of departure or when contact is made with the enemy, whichever comes first.

Movement to the Assembly Area. The squad and fire teams adopt formation and movement techniques en route to the assembly area, usually as part of a larger element, and influenced by—

- The geographic relationship between units.
- The anticipated reaction of friendly units to enemy contact.

- The level of security desired.
- The posture of friendly forces for attack or defense.
- The orientation of the preponderance of subordinate units' weapons systems.

Movement Formations. Movement formations are the systematic arrangement of elements that depict the general configuration of a unit on the ground. The formation adopted during movement to the assembly area will usually be one or a combination of the following, dependent on the considerations stated above:

- Column.
- Line.
- Echelon.
- Wedge.
- Vee.

Formations do not demand restrictive precision; squads must retain the flexibility needed to vary their formations to the situation. The use of formations allows Marines to execute immediate action drills more quickly and gives them the reassurance that their leaders and fire team members are in the expected positions and performing the correct tasks. For more information on squad formations, refer to chapter 1.

Movement Techniques. Regardless of which formation is adopted during movement to the assembly area, one or more of the following three movement techniques will be used: traveling, traveling overwatch, or bounding overwatch. They refer to the distance between individual Marines, teams, and squads based on METT-T and any other factors that affect command and control. The selection of a movement technique is usually based on the probability of enemy contact and the need for speed (see table 3-2).

Table 3-2. Movement Techniques and Characteristics.

Movement Technique	When Typically Used	Characteristics			
		Control	Dispersion	Speed	Security
Traveling	Contact not likely	More	Less	Fastest	Least
Traveling Overwatch	Contact possible	Less	More	Slower	More
Bounding Overwatch	Contact expected	Most	Most	Slowest	Most

Leaders should consider control, dispersion, speed, and security when choosing one of the following movement techniques:

- *Traveling*. Traveling is used primarily when contact with the enemy is not likely and the need for speed is required.
- *Traveling overwatch*. Traveling overwatch is used when contact is possible. If the squad has crew-served weapons attached, they should be positioned in close proximity to the squad leader and under the squad leader's control for quick employment.

- *Bounding overwatch*. Bounding overwatch is used when contact is expected, when the squad leader feels the enemy is near (based on indicators such as movement, noise, reflection, trash, fresh tracks, or even a hunch) or when a large open danger area must be crossed. The lead fire team provides overwatch first. Marines in the overwatch team scan for signs of enemy presence. The squad leader usually stays with the overwatch team. The trail fire team bounds and signals the squad leader when the team completes its bound and is prepared to overwatch the movement of the other team.

Fire team leaders must know which team the squad leader will be located with. The overwatch fire team leader must know the route and destination of the bounding fire team. The bounding fire team leader must know their team's destination and route, possible enemy locations, and actions to take when they arrive (based on unit SOP). They must also know where the overwatch team will be and how they will receive instructions. Available cover and concealment along the bounding team's route dictate how its Marines move.

Movement Elements. As an element of one of the aforementioned movement formations, the squad may be assigned to one of the following roles:

- Main body.
- Advance guard.
- Rear guard.
- Flank guard.

When the squad moves as part of the main body, the squad leader's primary duties involve the supervision of march discipline within the squad.

The squad may also be assigned to the forward security element of the advance guard. The advance guard precedes the main body along the axis of advance (i.e., the unit's general direction of movement). The distance between the main body and advance guard is determined by the commander and is usually based on METT-T. Its mission is to prevent an enemy in the route march vicinity from surprising the unit, and to prevent any undue delay by reducing obstacles (using attached combat engineers), developing the situation, or conducting other activities designed to cover the deployment of the main body. Possible ambush sites such as stream crossings, road junctions, small villages, and defiles are thoroughly probed by the advance guard.

The squad leader determines squad formations, usually according to terrain or potential for enemy contact. Generally, the squad uses a wedge or open column formation (see figure 3-1). If the squad is advancing in the wedge formation, the lead fire team moves on the edges of the road or trail. The two fire teams in the rear march off the road or trail, with one team on either side of it. When the road or trail is bound by thick vegetation or there is a need for haste, the formation of the point is usually a column. The fire teams may also be in column formation and advance along alternate sides of the road or trail. In any event, it is the squad leader's responsibility to change the formation when the need arises.

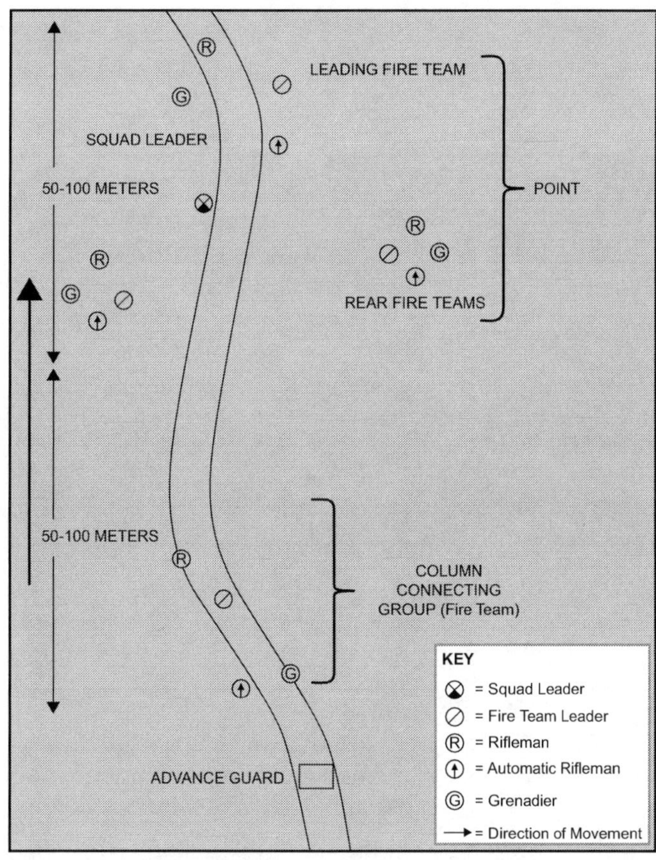

Figure 3-1. Squad as Security Force of an Advance Guard.

The squad leader assigns each fire team a sector of observation (see figure 3-2) and the fire team leaders assign each individual a sector of observation. Individual sectors of observation should overlap so there are no gaps in the squad or fire team sectors of observation. This ensures the all-round observation essential for the proper security of the point element.

Squad leaders generally position themselves just to the rear of their lead fire team or the position from which they can best control their squad. They are far enough to the rear to avoid being pinned down by the initial burst of any enemy fire, and yet far enough forward for continuous reconnaissance, which enables formulating an estimate of the situation and making decisions in minimal time. The squad leader and the fire team leaders must continually check to see that all members of the squad are alert and vigilant at all times. Weapons are carried ready for immediate use. Whenever possible, the point Marine uses hand-and-arm signals for communication.

The squad leader reports enemy contact (e.g., enemy units, obstacles, explosive hazards, or UASs) to the advance guard commander and describes the enemy situation and action being taken. If the enemy resistance is weak in comparison to the strength of the security force, the squad leader initiates a plan to close with and destroy the enemy. If the enemy resistance is greater than the strength of the security force, the squad attacks in a manner that forces the enemy to open fire and disclose its disposition and strength. Such aggressive action significantly assists the advance guard commander attain a precise estimate of the situation. When the squad makes visual contact with an enemy along the route of march but beyond its effective range, the advance guard

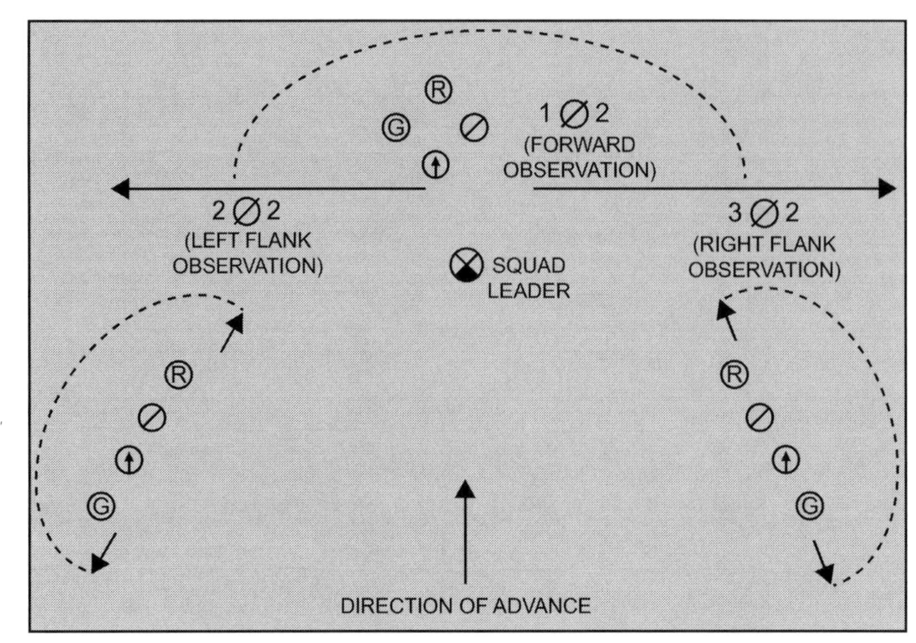

Figure 3-2. Fire Teams Sectors of Observation.

commander is notified and the advance continues until contact is made with the enemy. When the enemy is observed beyond effective range to a flank, the squad does not proceed to make contact with the enemy, but instead notifies the advance guard commander.

In the same manner that the advance guard may employ a squad forward, the rear guard may employ a squad to cover its rear. The squad providing rear security is arranged similar to that of the lead squad in the advance guard, but in reverse. The squad generally employs a vee or column formation, with the squad leader at the head of the rear-most fire team. This formation is easy to control, provides all-round security, favors fire and maneuver to the flanks, and allows adequate fire in all directions (see figure 3-3). The rear security force stops to fire only when enemy action threatens to interfere with the march. Observed enemy activity is reported to the rear guard commander. The rear security cannot expect reinforcements; it repels enemy attacks vigorously.

The missions, actions, and formations used by a squad when serving as the security force of a flank guard are the same as when the squad is acting as the security force of an advance guard.

Squads are often detailed as flank security elements. A flank guard may be ordered to move to and occupy an important terrain feature on the flank of the advance, or to move parallel to the column at a prescribed distance from it (i.e., the distance depending on the speed of the column), the terrain, and the enemy. When vehicles or air assault assets are available and terrain permits their use, it is highly desirable to provide the flank guard element with transportation.

When conducting dismounted movement parallel to the main body, the squad should adopt formations based upon considerations such as terrain, speed, and self-protection. In open terrain, a wedge formation is usually the best. In heavily wooded terrain, the squad might use the column formation. The lead fire team usually serves as the scouting element.

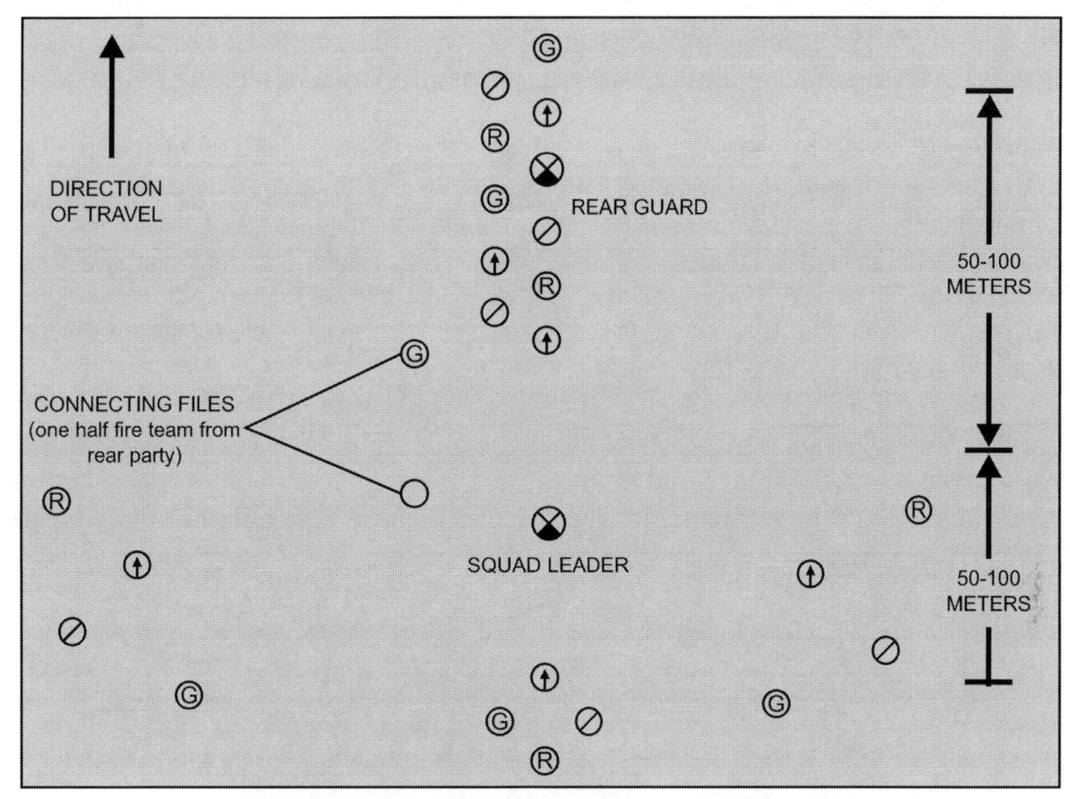

Figure 3-3. Squad as Security Force of Rear Guard.

The squad moves so as to prevent the enemy from placing effective small arms fire on the protected column. It investigates areas that are likely to conceal enemy elements or provide them good observation. The squad observes from commanding terrain, moves rapidly from point to point, and maintains positions between the protected column and possible enemy locations.

Enemy elements moving away from the main body are reported, but are not engaged unless otherwise directed. All other enemy forces within effective range are engaged immediately. If the enemy engages either the flank squad or the protected column, the squad determines the enemy's strength and disposition and reports this information promptly to the unit commander. The squad engages and repels any enemy attack until ordered to withdraw.

Overwatch for Halted Column. Tactical overwatch is used by a moving unit making a temporary halt. It is established by the advance, flank, and rear guards, who occupy critical terrain features to watch avenues of approach to the halted column; special attention is given to the flanks.

The mission of the overwatch units is to protect the halted column from surprise attack by the enemy. If attacked, overwatch units engage the enemy, allowing the column time to take up a defensive position from which to repel the attack.

A squad is often tasked to provide tactical overwatch. The platoon commander informs the squad leader of the situation, the overwatch position to be occupied, where and to whom to send reports of enemy activity, and the anticipated duration of the halt. Upon arrival at the prescribed location and making a hasty reconnaissance, the squad leader positions the fire teams where they can observe and defend all avenues of approach. Alert observation is ensured by establishing observation posts (OPs) and arranging for frequent reliefs (see figure 3-4). The squad does not abandon its overwatch position until it receives explicit orders to rejoin its unit.

Termination of the Tactical Movement. The tactical movement usually ends when the unit occupies its assembly area to prepare for an attack. However, the enemy situation may cause a unit to deploy immediately into tactical formations without occupying an assembly area.

Preparations in the Assembly Area. An assembly area is an area in which a unit is assembled preparatory to further action. Ideally, an assembly area should be located outside the effective range of enemy medium indirect fires and provide—

- Concealment from air or ground observation.
- Suitable entrances, exits, and internal routes.
- Space for dispersion from other assembly areas to preclude mutual interference as required.

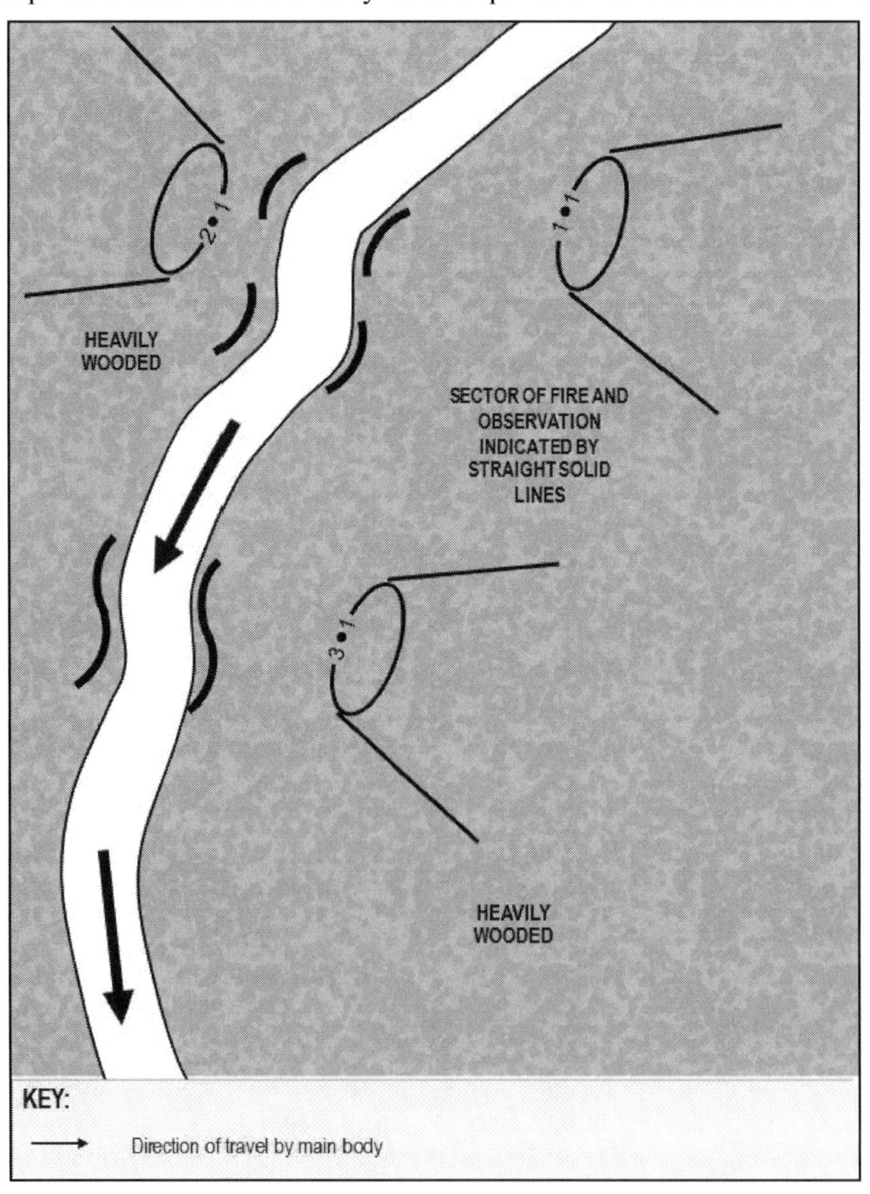

Figure 3-4. Squad Providing Tactical Overwatch.

- Cover from enemy direct fire.
- Ground conditions that support vehicle and individual movement.
- Observation of air and ground avenues into the assembly area.

Squads occupy the assembly area as part of a larger unit, usually the platoon. In the assembly area, squad leaders plan, direct, and supervise their squads' mission preparations for upcoming actions. The squad leader should utilize this time to conduct pre-combat checks, pre-combat inspections, rehearsals, and draw and issue extra ammunition or special equipment. Preparations conducted by the squad leader in the assembly area include the following:

- Complete the plan of attack. The fourth step in the troop leading procedures (complete the planning) requires the squad leader to complete the plan of attack, paying particular attention to actions on the objective. The squad is usually confined to two forms of maneuver in an attack—frontal or flanking. For more information on the forms of maneuver, see MCDP 1-0, *Marine Corps Operations.*
- Issue the order. Once the squad leader has completed the plan of attack, they issue the order. The squad leader should utilize the five-paragraph order format located in appendix C. The squad leader should detail how the squad fits into the company's and platoon's overall plan from the assembly area to the objective. Details should be given on all tactical control measures, the type of attack, routes, the attack position, line of departure, etc. (see figure 3-5). The order consists of the following:
  - Situation (with orientation).
  - Mission.
  - Execution.
  - Administration and Logistics.
  - Command and Signal.

Movement to the Line of Departure. The squad typically moves from the assembly area to the line of departure as part of the platoon's movement plan. This plan may direct the squad to move to an attack position to await orders to cross the line of departure. The squad and fire team leaders must know where they are to locate within the assigned attack position, which is the last position an attacking element occupies or passes through before crossing the line of departure.

Approach March. The squad leaves the assembly area and continues the movement toward the enemy in the approach march formation, usually as part of the platoon. The approach march formation (see figure 3-6) is used when enemy contact is imminent. The column establishes guards to the front, flanks, and rear, as appropriate. Elements within the column may be fully or partially deployed in the attack formation. The advance is usually made by bounds, stopping on easily recognizable terrain features to coordinate the advance and provide overwatch. During the approach march, the squad and fire teams take maximum advantage of cover and concealment along the route.

The squad's initial approach march formations may be dictated by the platoon commander. As the march progresses, the squad leader may order formation changes based upon terrain, assigned sectors, and the likelihood of enemy contact. The platoon commander may also designate a base squad to maintain direction, control the rate of march, and for other squads to use as a guide.

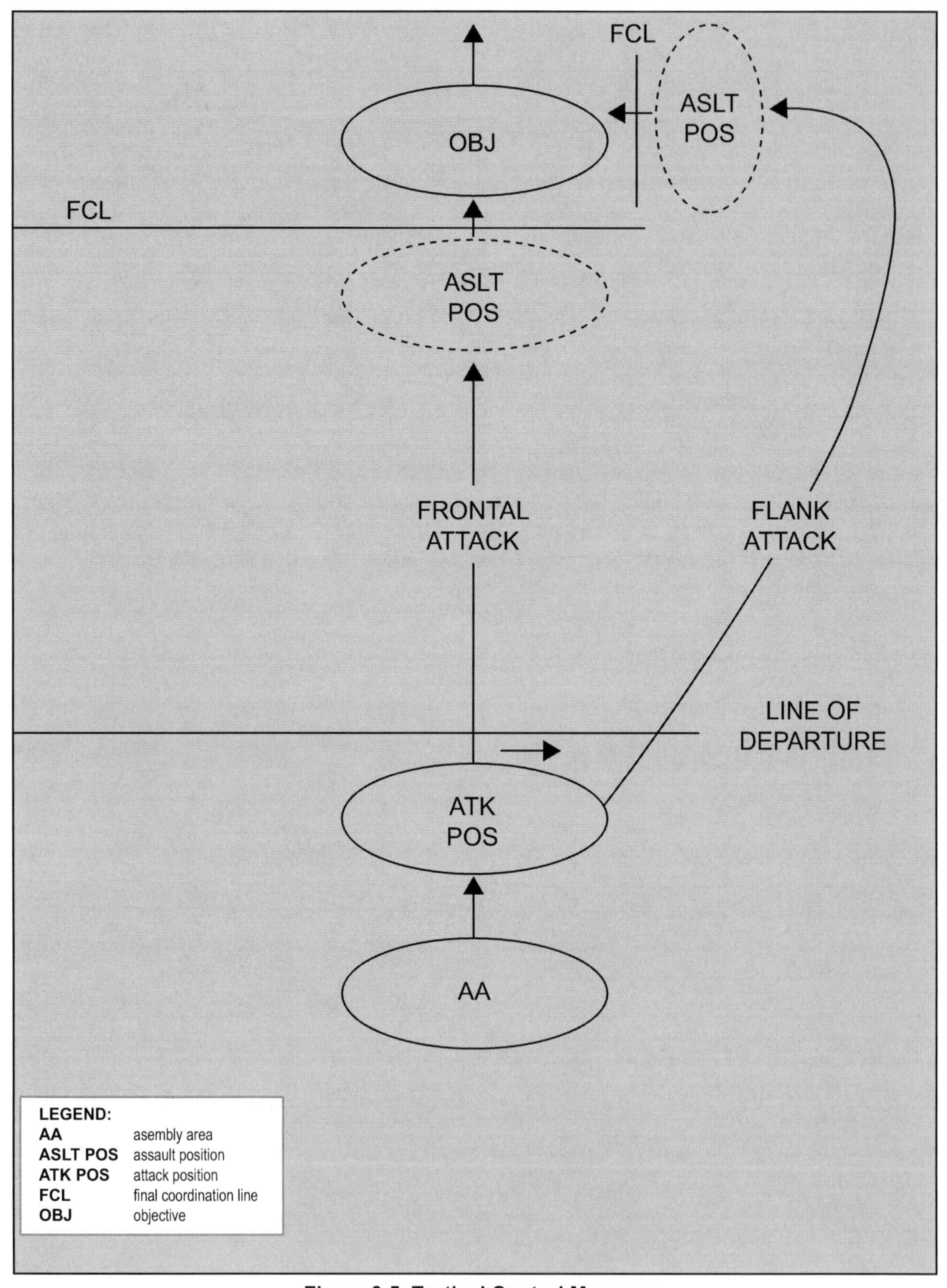

Figure 3-5. Tactical Control Measures.

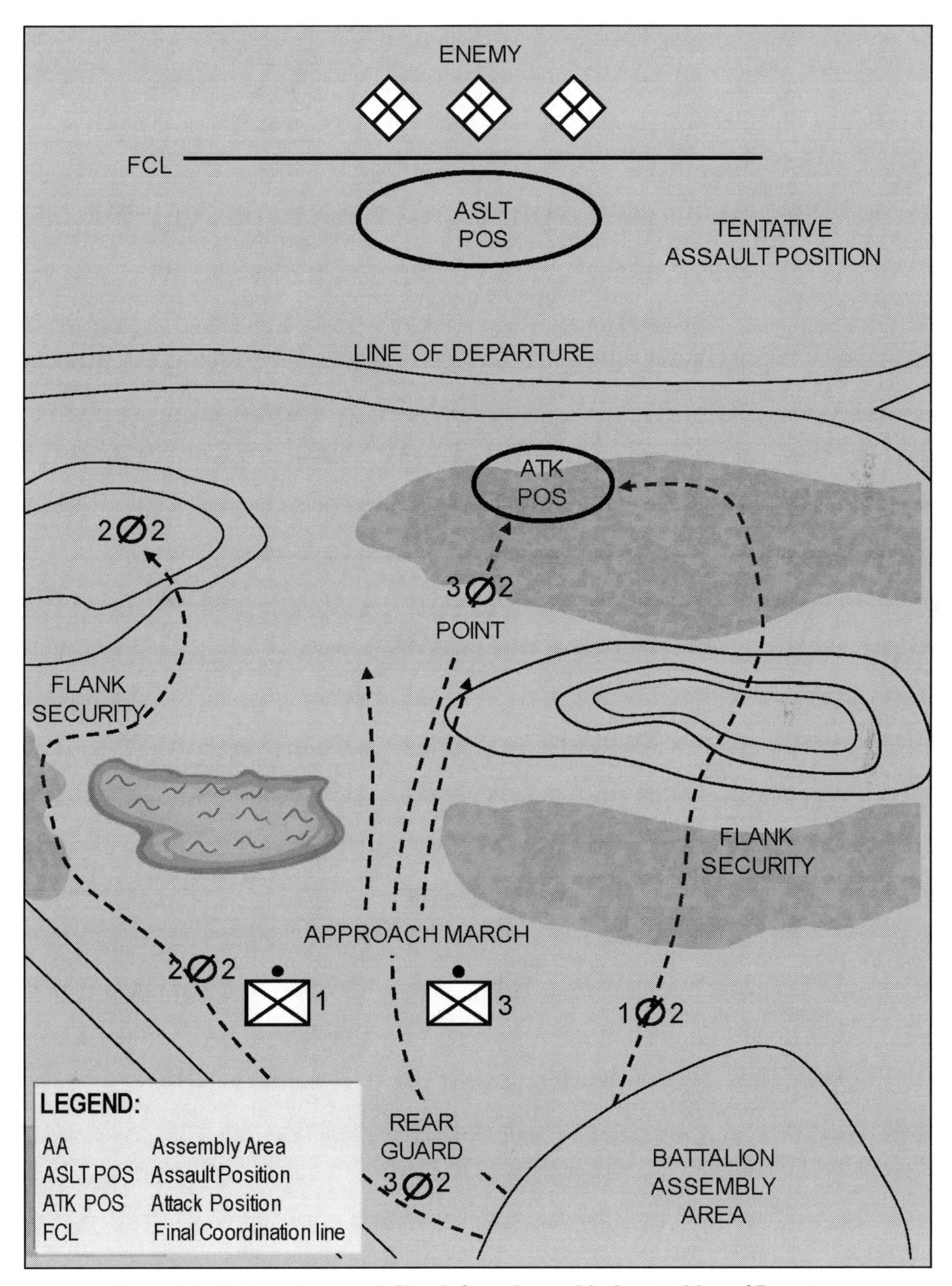

Figure 3-6. Platoon Approach March from Assembly Area to Line of Departure.

The squad leader regulates the squad's advance on the movement of the platoon's base squad, or if they are the base squad, they advance as directed by the platoon commander. As the squad leader moves, they study the ground to the front and flanks in order to take advantage of cover and concealment and to control the movement of the fire teams. They also maintain direction and make minor deviations to take advantage of better terrain.

Scouting Fire Team. When a company or platoon in the approach march is not preceded by other friendly forces, it uses its own scouting elements. The scouting element usually consists of one fire team; however, an entire squad may be used. A fire team used as a scouting element is called a scouting fire team. It is controlled by the platoon commander with assistance from the squad leader. A scouting fire team moves aggressively to cover the front of the advancing platoon, attempting to force the enemy to disclose their positions. The formations generally used by a scouting fire team are based on the terrain and the probability of enemy contact. Scouting fire teams are covered by the platoon, or when the platoon is masked, it covers its own advance. The fire team leader watches constantly for signals from the platoon commander, trying to remain in visual contact at all times. The distance that the scouting fire team moves ahead of the platoon is typically the limit of visibility and varies with the terrain. In open terrain, the platoon commander usually directs the scouting fire team to move by bounds along a succession of objectives. For more information, see MCTP 3-01A, *Scouting and Patrolling.*

When a scouting fire team is fired upon, the individuals immediately take cover, locate targets, and return fire. The scouting fire team leader then determines—

- The enemy location.
- The size of the enemy position (i.e., left and right flanks).
- The types of positions.
- The enemy's strength.
- The enemy's weapons systems.

The platoon commander contacts the leader of the scouting fire team to obtain as much information as possible, then returns control of the scouting fire team to the squad leader. Usually, the platoon commander brings up the remainder of the squads, sets up a base of fire, and assaults the enemy position. However, the decision may be made to isolate the enemy position and bypass it, leaving it to be dealt with by follow-on forces according to the bypass criteria.

Attack Position. The attack position is the last position an attacking force occupies or passes through before crossing the line of departure. It is the location where final coordination, last minute preparations, and deployment into initial tactical formations (if not already accomplished) are conducted. When all preparations for the attack are completed in the assembly area, there should be no delay when passing through the attack position.

Conduct

Line of Departure to the Assault Position. The squad, moving as part of the platoon, crosses the line of departure towards the assault position. The squad's movement is usually covered by the company's fire support plan or other shaping actions. The squad accomplishes its movement or maneuver utilizing the fires provided by the support element. Once the squad reaches the assault position (i.e., the last covered and concealed position before the objective), it should immediately deploy into its assault formation and conduct final coordination.

Assault Position. The assault position is a covered and concealed position short of the objective from which final preparations are made to assault the objective. The assault position should be located as close to the objective as the assaulting elements can move by fire and maneuver without sustaining casualties or masking covering fires, both direct (i.e., base of fire) and indirect (i.e., artillery and mortar). The assault position should be easily recognizable on the ground and should offer cover and concealment to the attacking force. Here, the final steps are taken to ensure a coordinated assault. Only a minimal amount of time should be spent in this position to preclude the enemy from fixing the assault element in place.

When the squad reaches the assault position, the squad leader, fire team leaders, and squad members must quickly make final preparations for the assault. Unit leaders issue last minute instructions to their elements.

Warning

Units should not delay in the assault position.

Assault Position through the Primary Objective. The primary object in advancing the attack by fire and maneuver and/or fire and movement is to get part of or the entire attacking unit in position to assault the enemy. The position from which the final assault commences is the assault position. As the squads close on the objective, the effects from supporting fires increase in intensity. To avoid possible fratricide, cease or shift supporting fires just prior to reaching the objective. The control measure used to coordinate this is the final coordination line. Both the assault position and final coordination line are crucial to the assault.

Momentum. Momentum in the assault is critical, and the assault position is not a rest point—if any time is spent there at all, it is minimal. Final preparations conducted there include things such as ensuring the readiness of weapons and special equipment, reorganization due to combat losses, or adjustments to the attacking force's disposition.

Final Coordination Line. The final coordination line is used to coordinate the ceasing and shifting of supporting fires and the final deployment of the assault echelon in preparation for launching an assault against an enemy position. Final adjustments to supporting fires necessary to reflect the actual versus the planned tactical situation are made prior to crossing this line. It should be easily recognizable on the ground.

Squad Employment. The squad is typically employed as part of the rifle platoon and will usually be assigned a mission as a base of fire or as a maneuver element. Thus, when operating as part of the platoon, a squad assigned as the maneuver element executes fire and movement or fire and maneuver. For example, a squad is required to fire and maneuver when given a role such as point squad, flank patrol, or flank guard during a movement to contact and enemy contact is made. The organization of the rifle squad into three fire teams provides the squad leader the ability to execute fire and maneuver with one or two fire teams employed as the base of fire and one or two fire teams as the maneuver element.

The base of fire covers the maneuver element's advance toward the enemy position by engaging all known or suspected targets. Upon opening fire, the base of fire seeks to gain fire superiority over the enemy. Fire superiority is gained by subjecting the enemy to fire of such accuracy and volume that the enemy fire ceases or becomes ineffective. The squad may be assisted with this mission with the augmentation of crew-served weapons.

The mission of the squad as part of the maneuver element is to close with and destroy or capture the enemy. The squad advances and assaults under the covering fire of the base of fire element. The maneuver element takes maximum advantage of all cover and concealment. Depending upon the terrain and the effectiveness of the covering fire, the maneuver element advances by team movement. Within the team, Marines advance by fire and movement, employing rushes or crawling as necessary. Regardless of how it moves, the squad must continue to advance. If the terrain permits, the squad may be able to move forward under cover and concealment to positions within hand grenade range of the enemy.

Method of Advance. A rifle squad has four methods by which it may conduct fire and movement. The squad may move as a unit in a series of squad rushes, as fire teams in a series of alternating fire team rushes, in buddy rushes, or the members of the squad may move forward by individual rushes. The volume of the enemy's fire, terrain, and proximity will determine which method the squad uses.

Control of the Squad. Fire team leaders initiate the action directed by the squad leader. In the attack, fire team leaders serve as fighter-leaders, controlling their fire teams, primarily by example. Fire team members base their actions on the actions of their fire team leader. Throughout the attack, fire team leaders exercise such positive control as is necessary to ensure that their fire teams function as directed. The squad leader locates in a place that offers the best control and influence. In controlling the squad when under enemy fire, the squad leader takes into account the fact that the battlefield is a noisy and confusing place. If enemy fire is light, the squad leader may be able to control the fire team leaders by voice or through intra-platoon radios. As the volume of enemy fire increases, this type of control becomes impossible, and whistles or hand-and-arm signals may have to be used (see appendix B). In this situation, the squad leader must rely on the skill and initiative of the fire team leaders to carry out the instructions given to them previously. To maintain control of the squad under heavy enemy fire, squad leaders position themselves near the fire team leader of the designated base fire team. By regulating the actions of the base fire team leader, the squad leader retains control of the squad. The base fire team leader controls the actions of their fire team; the other fire team leaders base their actions on those of the base fire team. This type of control must be practiced and perfected in training if the squad is to be effective in combat.

The base fire team is used by the squad leader to control the direction, position, and rate of movement of the squad. It is not intended that the other fire teams maintain rigid positions in relation to the base fire team; the base fire team is used as a general guide. If another fire team can move forward more rapidly than the base fire team, it should do so. For example, if the base fire team is receiving enemy fire, but the terrain in front of another fire team provides cover from enemy fire, the latter team should move rapidly forward to a position where they can deliver fire on the enemy. Covering the base fire team's movement by fire takes pressure off them and permits them to move forward. Once the base fire team comes generally abreast, the other fire teams can then resume fire and movement.

Squad in the Assault. The assault must be launched close under the covering fires and begin when the leading assault elements have advanced as close to the enemy as possible without moving into friendly covering fires. The assault is started either on the order or signal of the platoon commander, or on the initiative of the squad or fire team leader. Supporting weapons cover the assault by firing on adjacent or deeper enemy elements. The assault is launched aggressively and vigorously **immediately** upon shifting or ceasing the covering fires on the objective. The squad should advance rapidly and aggressively from the assault position using assault fire techniques.

Assault fire is designed to keep the enemy suppressed once covering fires are lifted by fixing the defenders in their fighting positions. Assault fire permits the assaulting squad to close to within hand grenade range of the enemy position without sustaining heavy casualties from enemy small arms fire. The assault is made as rapidly as possible based on the ability of individuals to deliver a heavy volume of well-aimed fire. The speed of the assault will be governed by the slope and condition of the terrain, visibility, and physical condition of the squad members. Assault fire is characterized by violence, volume, and accuracy. Assault fire is designed to kill or suppress the enemy until the assault element can overrun the position.

Decentralization of Command. Squad leaders are fighter-leaders during the assault. They position themselves where they can move rapidly to enforce continuity of fire, maintain alignment and momentum, and keep the assault moving forward aggressively, concentrating on the actions of their squad rather than engaging the enemy with their individual weapons. If the assaulting squad is faced with light enemy opposition, it may be possible for the squad leader to retain control of the unit by maintaining the assault line and sweep through the objective. However, when enemy opposition is heavy, maintaining the squad skirmish line may not be possible. When assaulting an organized position, the squad attack may break into a series of separate actions throughout the depth of the enemy position. Controlling the squad under these conditions is difficult—the squad leader must rely on the skill and initiative of the fire team leaders and individual Marines to carry out the assigned mission. Each fire team leader and individual Marine must take the initiative to use individual weapons, grenades, and other ordnance to maximum effect. The first fire team to gain a foothold on the enemy position supports the remainder of the squad in seizing the objective.

Consolidation and Reorganization

The purpose of consolidation and reorganization is to rapidly and efficiently prepare the attacking force for future action during the exploitation step. Both actions are mutually supporting, but not automatically linked—while squads should consolidate, they do not always need to reorganize.

Consolidation. Consolidation pertains to those measures taken to organize and strengthen a newly captured position to use against the enemy. Initially, assault and support elements assume a defensive posture to repel possible enemy counterattacks. The squad should initiate all immediate actions to occupy and defend their assigned sector. Consolidation actions include—

- Maintaining contact with the enemy.
- Establishing security and conducting reconnaissance.
- Establishing a defense against possible enemy counterattacks and indirect fires.
- Receiving updates on enemy dispositions, fire support, and IO plans.
- Preparing for additional follow-on missions.

Enemy Counterattack. The primary concern of any leader once the enemy has been driven from the objective is to retain control of the objective. If the enemy positioned troops to defend the objective and offered stiff resistance, it is reasonable to assume that they may attack to reclaim it. It is safe to say that it is not a question of whether or not the enemy will counterattack, but when. The enemy knows that their chance of success is better if they counterattack quickly before there is time to build a strong defense. By striking quickly, the enemy will not give the defenders time to bring up reinforcements. The wise Marine will expect an enemy counterattack even before the last enemy positions on the objective have been neutralized. Preparations to repel the counterattack must commence immediately after taking the enemy position.

Hasty Defense. In positioning the fire teams in a hasty defense, there is not sufficient time to prepare standard fighting positions. The squad must use natural depressions, shell craters, or old enemy positions if available, and quickly improve them to provide minimum adequate cover, as the enemy will use all direct and indirect fire weapons available to wage a counterattack. The priority must be to effectively defend the squad's assigned sector by fire, and to get the squad under cover quickly, not perfectly. Because of the rapid tempo of events, the full attention of the squad and fire team leaders must be dedicated to preparing the hasty defense. Fire team sectors of fire are designated and principal directions of fire for automatic rifles are assigned. The squad leader must ensure that each fire team leader takes the initiative to ensure that their team's sector of fire is interlocked with that of the adjacent teams. If the squad leader or a fire team leader has become a casualty, the next senior Marine must quickly assume control and carry out the necessary tasks.

When enemy resistance on the objective is heavy, the fire team leader should remain close to the automatic rifleman and attempt to get automatic riflemen through the objective and integrated into the hasty defense as quickly as possible. The firepower of the automatic rifle is critical to the hasty defense. When the squad assault breaks up into a series of individual combat, the squad leader's task of building a hasty defense is made even more difficult because the squad's organizational integrity has temporarily broken down. The squad and fire team leaders are now faced with the task of building a hasty defense without having the usual control of the unit. In this situation, the squad leader proceeds throughout the objective, gathering as many members of the squad as possible and positioning them to cover the assigned defensive sector. Due to the confusion of the battle for the objective, there will be individual Marines in the squad sector who have been separated from their fire teams. The squad leader should take charge and temporarily assign these Marines to fire teams already in position, or if necessary, form temporary fire teams. If a separated Marine is wandering around looking for their fire team, that firepower will be lost to the unit if the enemy counterattacks. During consolidation, the primary task is holding the objective, not reforming the squad. Unit leaders must make their presence felt when building the hasty defense to assure each Marine that they are not alone in resisting the enemy counterattack.

Reorganization. Reorganization is shifting internal resources within a degraded unit to increase its level of combat effectiveness. It often takes place with consolidation, though the two are not automatically linked. If required to reorganize, the squad leader should accomplish the following:

- Ensure positive communications are restored (if necessary).
- Replace fire team leaders and automatic riflemen who have become casualties.
- Redistribute arms and ammunition, and replenish supplies.

- Conduct casualty evacuation (CASEVAC).
- Notify the platoon commander of the squad's situation (e.g., submit a casualty report and ammunition status).
- Deliver any enemy detainees to the platoon commander.
- Conduct site exploitation (see appendix E for more information).
- Determine the status of the units on the squad's flanks.

Exploitation

Exploitation typically begins immediately after or in conjunction with the consolidation and reorganization step. It is a continuation of the attack aimed at destroying the enemy's ability to conduct an orderly withdrawal or organize a defense. The squad will usually participate in exploitation as part of a larger force (i.e., platoon or company). However, the squad should be prepared to exploit tactical success at the local level within the commander's intent. For more information on exploitation, see MCDP 1-0.

Pursuit

The objective of the pursuit step is the total destruction of the enemy force. The squad typically takes part in a pursuit as part of a larger force (i.e., company or platoon), depending on the size of the enemy force and the available transportation assets, which enable the pursuit force to close with and destroy the remnants of the enemy force. A pursuit typically follows a successful exploitation and is designed to prevent a fleeing enemy from escaping and to destroy them. For more information on pursuit, see MCDP 1-0.

Attack is often the preferred action during continuous operations, and is typically conducted hastily based on unforeseen opportunities. The squad participate in hasty attacks as part of a larger unit during movement to contact, as part of a defense, or whenever the commander determines that the enemy is vulnerable. Hasty attacks are used to—

- Exploit tactical opportunities.
- Maintain momentum.
- Prevent the enemy from reorganizing.
- Gain a favorable position.

Because the purpose of the hastily conducted attack is to maintain momentum and take advantage of an unfavorable enemy situation, there is usually very little time to prepare for it. Therefore, it is usually executed with the resources that are immediately available. The objective of the hasty attack is to maintain pressure on the enemy and keeping them off balance, denying them the ability to react in a cohesive manner. An attack in this manner, focused on agility and surprise, can cause synchronization problems for leaders. Therefore, squad leaders should strive to—

- Maximize use of standard formations.
- Use well-rehearsed and thoroughly understood battle and immediate action drills.
- Use company and platoon SOPs.

ATTACKS IN LIMITED VISIBILITY

An attack might be made during periods of limited visibility to gain surprise, to maintain pressure, to exploit a success in continuation of daylight operations, to seize terrain for subsequent operations, or to avoid heavy losses by using the concealment afforded by darkness. As used here, the term limited visibility applies to periods of darkness, not to short durations of reduced visibility caused by rain, snow, fog, or battlefield obscuration.

The squad equipped with night vision devices (NVDs) and thermal imaging devices can conduct a limited visibility attack in a manner similar to a daylight attack. The fundamentals still apply, though more emphasis is placed on small unit leaders.

Effective use of NVDs and thermal imaging devices can enhance the squad's ability to achieve surprise and cause panic in a lesser-equipped enemy. Night vision device enhancements allow Marines to see farther and with greater clarity, providing a clear advantage over the enemy. However, leaders cannot assume that we possess a significant technological superiority over our adversaries; to do so is a grave error.

Squad and fire team leaders have a greater ability to control fires during limited visibility. Leaders can designate targets and sectors of fire with greater precision using laser aiming devices attached to their individual weapons. The squad leader "designates" a target and directs Marines to concentrate fires on it, using the laser aiming devices attached to their individual weapons.

The squad leader and fire team leaders follow tactical SOPs to synchronize the employment of infrared illumination devices, target designators, and infrared lights. To reduce the risk to the assault element, the platoon leader may assign weapons control restrictions. Some of the following considerations may increase control during limited visibility attacks:

- The use of flares or smoke on the objective should be avoided.
- Only certain personnel with NVDs should be allowed to engage targets on the objective.
- A magnetic azimuth should be used to maintain direction.
- A base fire team should be assigned to maintain pace and direction to guide others.
- The intervals between Marines and fire teams should be reduced.

Preparation

Preparation for attacks during limited visibility is generally the same as for attacks in the daylight. Attacks conducted during limited visibility may tend to be more deliberate and possess significant psychological and tactical advantages over a less capable enemy. The squad leader follows the same orders format and is aided in preparation as outlined in appendices A and C. Squad leaders should place special emphasis on—

- Reconnaissance by squad and fire team leaders to locate both assigned control and terrain features during limited visibility orientation. If time permits, such reconnaissance should be conducted during three conditions of varying visibility—daylight, twilight, and night.

- Conducting rehearsals during both daylight and darkness. Rehearsals should include formations, audible and visual signals, and the actions of the squad from the assembly area to the objective, both with the aid of NVDs and without.
- Identifying and engaging targets.
- The ability to locate and evacuate casualties.
- Carrying only equipment deemed essential for successful accomplishment of the attack.
- The ability to conduct resupply.
- Protecting squad members' night vision, particularly for those not aided with NVDs.

Tactical Control Measures. The degree of visibility will determine the measures necessary to assure control, whether the attack is made with NVDs, illuminated, or non-illuminated. Terrain features used as tactical control measures may be marked by artificial means if not easily identifiable at night. At minimum, the following control measures are typically prescribed in a limited visibility attack (see figure 3-7):

- *Assembly area*. The assembly area may be nearer the line of departure than for a daylight attack.
- *Attack position*. The attack position should be in defilade, but need not offer as much concealment as in daylight. The area selected should be easy to move into and out of at night.
- *Line of departure*. The line of departure is a line established to coordinate attacking units when beginning the attack.
- *Release point*. Release points are clearly defined points on a route where units are released to the control of their respective leaders.
- *Probable line of deployment*. The probable line of deployment (PLD) is an easily recognized line selected on the ground where attacking units deploy in line formation prior to beginning a night attack.
- *Limit of advance*. A limit of advance is generally designated beyond the objective to cease the advance of attacking units. It should be a terrain feature that is easily recognizable in the dark (e.g., a stream, road, the edge of woods) and far enough beyond the objective to allow security elements space to operate.

Security Element. The squad may be employed as a security element during limited visibility attacks. These elements provide for security on the PLD, eliminate any enemy listening posts (LPs) and security elements, and prevent attacking forces from being ambushed while en route to the PLD. They may also act as guides to lead units forward from the release points to the PLD. The squad may accomplish this task using static positions or by conducting security patrols.

Conduct

Movement to the Probable Line of Deployment. The goal of the forward security element is to enable the main body to conduct movement to the PLD undetected.

The platoons move in column formation from the assembly area to their platoon release points. At the platoon release point, a platoon links up with its guides from the forward security element and continues to move along its respective route to the squad release point.

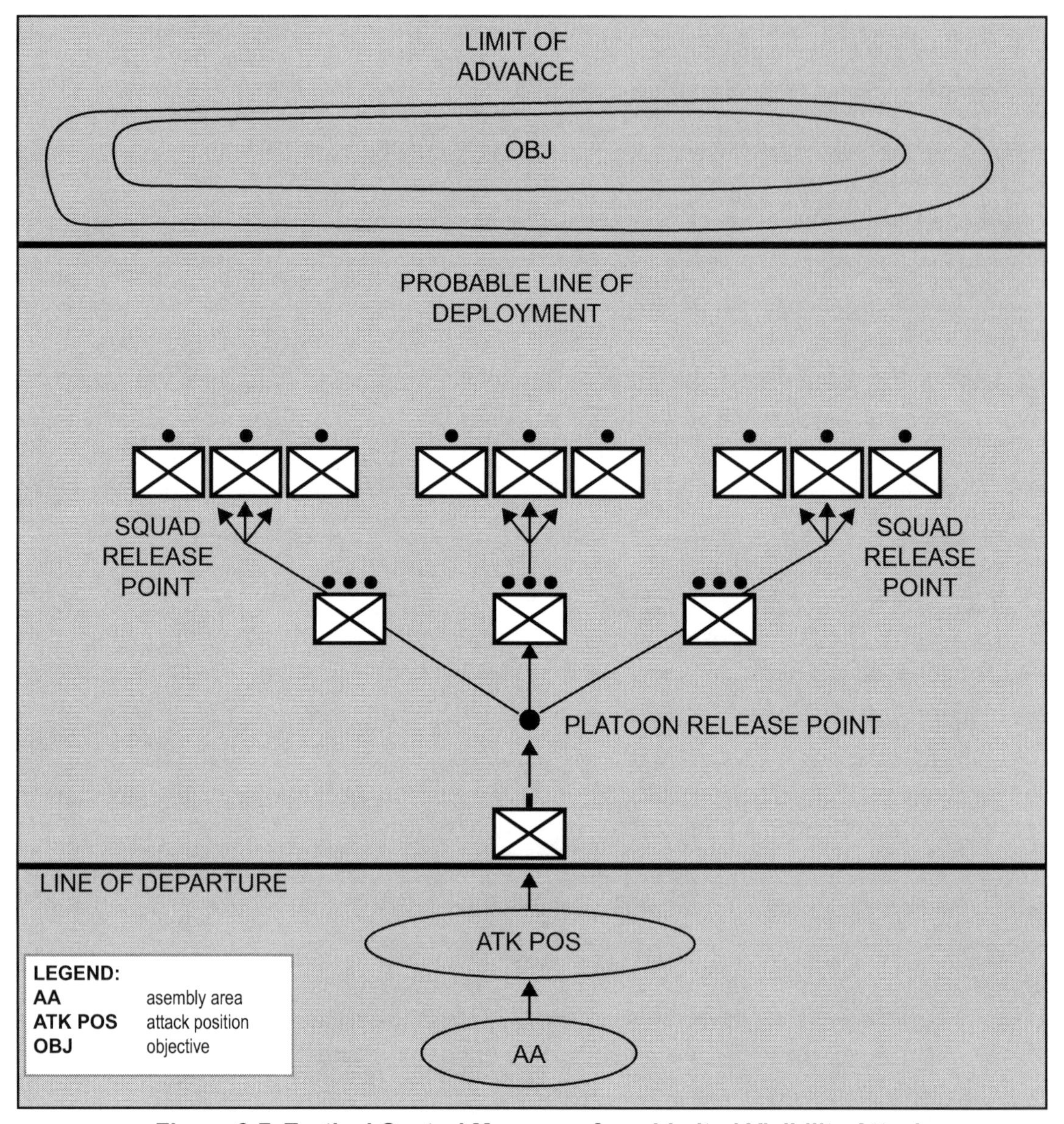

Figure 3-7. Tactical Control Measures for a Limited Visibility Attack.

Once the squad crosses the line of departure, movement to the PLD is continuous; the rate of advance should be slow enough to permit silent movement.

On arrival at the squad release point, the squads are released from the platoon formation to deploy on line at the PLD. The squad leader is typically the first member of the squad in the column. When the squad reaches the squad release point, the squad leader leads the column, sets the pace, and maintains the direction of movement. Members of the security patrols assist the squad leaders in positioning the squads on the PLD.

On order, the squad moves forward silently from the PLD, maintaining the squad line formation and guiding on the base squad.

> ***Note:*** If the attack is to be illuminated, the illumination starts on a pre-arranged signal from the attacking elements (usually after reaching the PLD).

Assault. The signal for the assault can take any form, but it must be simple and reliable. The importance of developing a high volume of accurate fire during the assault cannot be overemphasized; fire superiority must be established and maintained. The assault is conducted aggressively in the same manner as discussed for the daylight attack, and is carried forward to the forward military crest of the objective or to another prescribed limit short of the limit of advance.

Consolidation and Reorganization

When the objective has been seized, consolidation and reorganization are carried out in the same manner as for a daylight attack. The squad does not move or employ security elements forward of the limit of advance until ordered to prevent any instances of possible fratricide.

INFILTRATION

Infiltration is a form of maneuver in which friendly forces move through or into an area or territory occupied by either friendly or enemy troops or organizations. The movement is made either by small groups or by individuals at extended or irregular intervals.

The attacking force moves as individuals or small groups over, through, or around enemy positions without detection. The objective of infiltration is to move to a position of advantage into the enemy's rear area while only exposing small elements to enemy defensive fires (see figure 3-8).

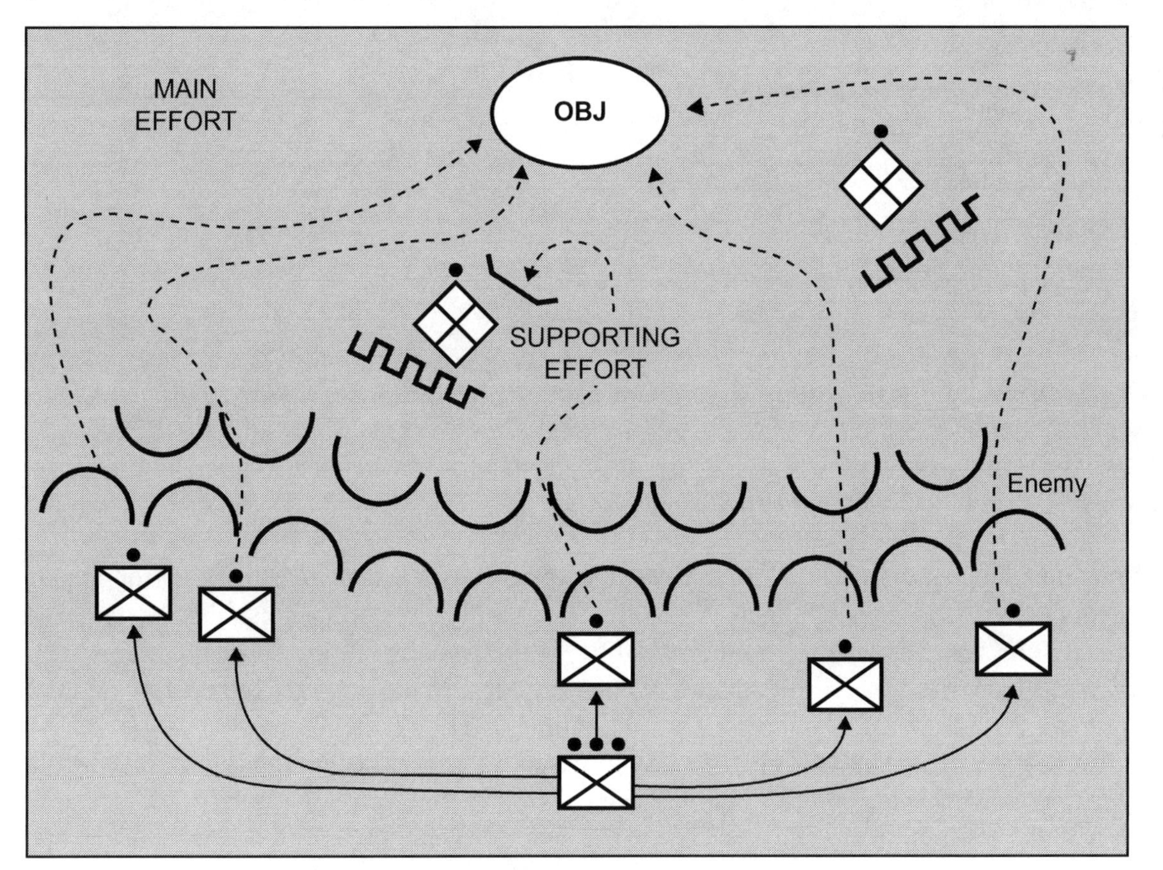

Figure 3-8. Platoon/Squad Infiltration Technique.

Although primarily offensive in nature, an infiltration can be conducted in conjunction with defensive or retrograde operations (i.e., exfiltration). For more information, see MCDP 1-0.

An infiltration may be used in conjunction with or in support of a unit conducting another form of maneuver (e.g., a squad infiltrates to establish a support by fire position to support a company conducting an envelopment). The squad as part of either the platoon or company may conduct an infiltration to—

- Attack enemy-held positions from an unexpected direction.
- Occupy a support by fire position to support an attack.
- Secure key terrain.
- Conduct ambushes and raids.
- Conduct a covert breach of an obstacle.

Planning and Preparation

Planning. A successful infiltration requires detailed and extensive planning. Combining the conditions of reduced and limited visibility with the synchronization of fires (i.e., both lethal and nonlethal) and maneuver require exacting detail. Upon the completion of planning, a detailed order is given. At minimum, the order should provide the squad leader with the following:

- Release points.
- Time of release.
- Point of departure.
- Time of infiltration.
- Infiltration lanes.
- Rendezvous point (i.e., primary and alternate).
- Time of rendezvous.
- Routes from rendezvous to attack positions.

The size of the infiltrating group will depend primarily on the need for control among infiltrating groups, the number and size of the gaps in the enemy's defense, and the amount of reduced or limited visibility (e.g., darkness, fog, snow, rain). Typically, units are broken down into infiltration groups no larger than platoon or squad size.

Preparation. Upon receipt of the order, the squad and fire team leaders follow the troop leading procedures as discussed in appendix A. While the squad leaders accomplish their troop leading steps, the fire team leaders prepare the squad for infiltration. Mission essential equipment is drawn, checked, and secured for silent movement. Each Marine prepares themselves and their equipment for the operation in accordance with unit's SOP. Whenever possible, each squad and fire team carries only the special equipment necessary to accomplish the mission of the infiltration force. This ensures the accomplishment of the mission in the event all groups do not successfully complete the infiltration.

After squad leaders issue their orders, rehearsals, pre-combat checks, and pre-combat inspections are conducted. Rehearsals should address the following at minimum:

- Passage of lines.
- Communications and signals.
- Actions at danger areas.
- Actions upon enemy contact.
- Actions to be taken at the rendezvous points and the objective.

Every Marine should be required to memorize the route to and the location of rendezvous points. Mission accomplishment relies primarily on the abilities of the small unit leaders. The planning and preparation must be as thorough and as detailed as time and facilities permit. Fires are planned by HHQ to create diversions and to protect and support the unit during the infiltration, in the rendezvous area, and during any subsequent attack, consolidation, or withdrawal.

Conduct of the Infiltration

The unit conducting the infiltration assembles the infiltration groups to the rear of friendly lines. The unit then moves forward until it reaches the release point. At the release point, the infiltration groups (i.e., platoons or squads) are released to their respective leaders. The squad moves by stealth to avoid detection, crossing the line of departure at the specified time, which is typically during periods of limited visibility. Artillery or mortar fires may be utilized as necessary to distract or deceive the enemy. The squads pass through the gaps in the enemy lines using the infiltration lanes. If detected, a squad should avoid becoming decisively engaged by withdrawing or moving around the enemy. Its speed of movement is limited by the requirement for stealth. Squads which are unable to reach their rendezvous points in time follow the previously briefed alternate plan.

At the rendezvous point, squads and platoons assemble and complete assault preparations. The first infiltration group to reach the rendezvous point secures it. The assembled force leaves the rendezvous point to assault the objective at the designated time in the same manner as in a limited visibility attack.

Attacking Fortified Positions

Fortifications are works emplaced to defend and reinforce a position. Time permitting, enemy defenders build bunkers and trenches, emplace protective obstacles, and position mutually supporting fortifications. Marines tasked with attacking prepared positions should expect to encounter any combination of planned enemy fires to include small arms, mortars, artillery, and antitank missiles, depending on the enemy's capabilities. If routed from their prepared positions, enemy troops may try to win them back by hasty counterattack. If the enemy is forced to withdraw, the squad should be prepared to encounter obstacles, ambushes, and other delaying tactics in an attempt to slow their pursuit.

Attacking fortified positions is time consuming, with greater emphasis placed on detailed planning. The degree of planning and preparation usually depends upon the character and size of the fortified position. Leaders must develop a scheme of maneuver that systematically reduces the area. Initially, these attacks should be limited in scope, focusing on individual positions and intermediate terrain objectives. Squad leaders must know and adhere to clear, concise bypass criteria and position destruction criteria and allocate forces to secure cleared enemy positions. Failure to retain cleared enemy positions may result in the enemy reoccupying them, isolating lead elements and ambushing follow-on elements.

Clearing enemy fortifications is clearly an infantry platoon function, because squads and platoons, particularly when augmented with engineers, are the best organized and equipped units in the Marine Corps for breaching protective obstacles. Infantry platoons and squads are also best prepared to assault prepared positions, such as bunkers and trench lines.

Leaders develop detailed plans for each fortification using a five-phase breaching technique—suppress, obscure, secure, reduce, and assault (see figure 3-9)—to integrate and synchronize fire support and maneuver assets. Although there are specific drills associated with types of fortifications, the assault of a fortified position is an operation, not a drill. Contingency plans must be made for the possibility of encountering previously undetected fortifications and clearing underground fortifications when encountered.

SOSRA
S – Suppress
O – Obscure
S – Secure
R – Reduce
A – Assault

Figure 3-9. Attacking Fortified Positions.

Breaching Protective Obstacles

Approaching fortifications usually requires penetrating extensive protective obstacles. Antipersonnel obstacles (i.e., explosive and non-explosive) are of particular concern to Marines. These include wire entanglements, trip flares, antipersonnel mines, field expedient devices, improvised explosive devices (IEDs), booby traps, rubble, CBRN, and any other type of obstacle created to prevent attacking forces from penetrating a position. Obstacles are usually covered by enemy fires close enough to the fortification for adequate surveillance during both daytime or reduced visibility. The following is an example of the drill a squad might conduct when executing a breach:

- The base of fire element commences suppression of the breach point (suppress).
- The squad leader and the breaching fire team (with attached combat engineers or infantry assault Marines) moves to the last covered and concealed position near the breach point.
- The squad leader confirms the breach point.
- The squad leader signals for shifting the suppressing element away from the breach point; the base of fire element continues to suppress enemy positions as required.
- Team 1 (the fire team leader and the automatic rifleman) occupies a position short of the obstacle to provide local security for team 2 (the grenadier, rifleman, and attachments).
- The squad leader and breaching fire team leader employ smoke grenades to provide obscuration of the breach point (obscure).
- Team 2 (the grenadier, rifleman, and attachments) moves to the breach point by moving in rushes or crawling.

- The squad leader occupies a position from which to best control the fire teams.
- Team 2 positions themselves to the right and left of the breach point near the protective obstacle, probing for mines and marking their path as they proceed (secure).
- Team 2 probes for mines and creates a breach; depending on the character of the obstacle, the breach may be executed by either explosive or mechanical means (e.g., wire cutters, bolt cutters).
- Once breached, team 1 and team 2 move to the far side of the obstacle, checking for mines and marking cleared lanes. They immediately take up covered and concealed positions to block any enemy movement toward the breach point and engage any identified or likely enemy positions.
- The squad leader remains at the entry point and calls forward the next fire team.
- Once the squad has secured a foothold, the squad leader reports to the platoon leader; the platoon follows the success seizing the foothold with the remainder of the platoon.

Attacking Bunkers

Bunkers are permanent; their location and orientation are fixed. Bunkers cannot be easily relocated or adjusted to meet changing situations. They are optimized for a particular direction and function. The worst thing an infantry squad can do is to approach a bunker in the way it was designed to defend against. Instead, the squad should approach the bunker from a direction it was least designed to defend against—the flank or rear.

Bunkers must have openings (i.e., doors, windows, apertures, or air vents). Structurally, the opening is the weakest part of the position and will be the first part of the structure to collapse if engaged. A single opening can only cover a limited sector, creating blind spots. To cover these blind spots, the defenders have to rely on mutually supporting positions or build additional openings. Mutual support should be disrupted, thereby enabling the squad to exploit blind spots.

Ideally, the squad is able to destroy bunkers with standoff weapons and high explosive munitions. However, the squad can assault the bunker with small arms and grenades when required. A fire team (i.e., two to four Marines) with grenades and/or demolitions and smoke grenades move forward under the cover of suppressive fires and obscuration from the rest of the squad and any other elements of the base of fire. When they reach a vulnerable point of the bunker, they destroy it or the personnel inside it with grenades or demolitions. All bunkers must be treated as if they contain living enemy personnel, even if no activity has been detected from within. Bunkers must be cleared systematically or the enemy may appear behind assault groups. To clear a bunker—

- The squad leader and the assault fire team move to the last covered and concealed position near the bunker's defenseless point (i.e., flank or rear).
- The squad leader shifts the base of fire away from the defenseless point.
- The base of fire continues to suppress the position and adjacent enemy positions as required.
- Team 1 (the fire team leader and the automatic rifleman) remain in a position short of the bunker to add suppressive fires for team 2 (the grenadier and rifleman).
- Team 2 moves to the defenseless point by moving in rushes or crawling.
- One Marine takes up a covered position near the bunker's exit.
- The other Marine shouts, *Frag out!* and throws it through an aperture.

- After the grenade detonates, the Marine covering the exit enters and clears the bunker.
- Simultaneously, the second Marine moves into the bunker to assist and gain cover.
- Both Marines halt at a point of domination and take up positions to block any enemy movement toward their position.
- Team 1 moves to join team 2.
- The fire team leader inspects the bunker, marks it, and signals the squad leader.
- The squad leader consolidates, conducts initial tactical site exploitation, and prepares to continue the mission. The squad may be tasked to support another squad's attack by fire or continue its attack on another objective.

Assaulting Trench Systems

Trenches are typically dug to connect fighting positions. They are typically dug in a zig-zag fashion to prevent attackers from firing down a long axis if they get into the trench and to reduce the effectiveness of high explosive munitions. Trenches may also contain shallow turns, intersections with other trenches, firing ports, overhead cover, tunnel systems, and bunkers. Bunkers are usually oriented outside the trench, but may also have apertures allowing the ability to provide protective fires into the trench.

Trenches provide defenders with frontal cover, allowing them the ability to relocate without the threat from the squad's direct fires. However, without overhead cover, trenches are vulnerable to the effects of high trajectory munitions like grenades, grenade launchers, plunging machine gun fire, mortars, and artillery. These weapons systems should be employed to gain and maintain fire superiority over enemy defenders in the trench.

Trenches' confined nature, extensive preparations, and limited ability to integrate combined arms fires against make clearing them hazardous for even the best trained infantry. In some instances, engineer equipment or tanks may be used to bury the defenders. However, since this is not always possible, infantry units must often move in and clear trenches.

Entering the Trench Line. To assault an enemy trench system, a squad conducts the following:

- The squad leader and the assault fire team move to the last covered and concealed position near the confirmed entry point.
- The squad leader shifts the base of fire away from the entry point.
- The base of fire continues to suppress the trench and adjacent enemy positions as required.
- Team 1 (the fire team leader and automatic rifleman) occupies a position short of the trench, adding suppressive fires for the initial entry.
- The squad leader and team 2 (the grenadier and rifleman) move to the entry point in rushes or by crawling; the squad leader takes a position from which best to control the fire teams).
- Team 2 positions itself parallel to the edge of the trench and get on their backs.
- On the squad leader's command, *Grenades!* team 2 Marines prepare grenades, shout, *Frag out!* and throw the grenades into the trench.
- Once both grenades detonate, the Marines move into the trench, assume a back-to-back stance, and engage all enemy personnel or suspected enemy positions.

- Both Marines move in opposite directions down the trench, continuing until they reach the first corner or intersection.
- Both Marines take up positions to block any enemy movement toward the entry point.
- Simultaneously, team 1 moves to and enters the trench, joining team 2. The squad leader directs them to one of the secured corners or intersections to relieve the Marine from team 2, who then rejoins their team member at the opposite end of the foothold.
- At the same time, the squad leader enters the trench and secures the entry point.
- The squad leader remains at the entry point, marks it, and calls the next fire team to the trench.
- Once the squad has secured a foothold, the squad leader reports to the platoon commander, *Foothold secure*. The platoon commander follows with the remainder of the platoon.

The platoon commander or a designated subordinate must move into the trench as soon as possible to control the tempo, specifically the movement of the lead assault element and the movement of follow-on forces. Leaders must resist the temptation to move the entire unit into the trench. Follow-on forces should be committed as elements reach their objectives and reinforcements are required.

The assault element (i.e., squad or platoon) should maintain fire team organization to clear the trench (Marines should be designated number one, number two, number three, etc.). Each team is armed with at least one automatic rifle, and all Marines are armed with multiple hand grenades.

The positioning within the fire teams is rotational, so the Marines must be rehearsed in each position. The responsibilities for each position are as follows:

- The number one Marine is responsible for assaulting down the trench using well-aimed, effective fire and throwing grenades around pivot points in the trench line or into weapon emplacements.
- The number two Marine follows the number one Marine closely enough to support them, but not so closely that both would be suppressed if the enemy gained local fire superiority.
- The number three and four Marines follow the number two Marine and prepare to move forward when the positions rotate.

While the initial fire team may rotate by event, the squad leader directs the rotation of the fire teams within the squad as ammunition becomes low in the leading team, casualties occur, or as the situation dictates. Since this drill should be SOP, fire teams may be reconstituted as needed from the remaining members of the squad. The platoon leader controls the rotation between squads using the same considerations as the squad leaders.

Clearing the Trench Line. The lead fire team of the initial assault squad moves out past the security of the support element and executes the trench clearing drill. The number one Marine (followed by the rest of the Marines in the fire team) maintains the advance until arriving at a pivot point, junction point, or weapon emplacement in the trench and conducts the following:

- The Marine alerts the rest of the team by yelling out, *Position!* or, *Junction!* and prepares a hand grenade.

- The number two Marine immediately moves forward near the number one Marine and provides covering fire until the grenade can be thrown around the corner of the pivot point.
- The number three and four Marines move forward to the point previously occupied by number two and prepare for commitment.

Anytime the number one Marine encounters a junction in the trench, the squad leader or a subordinate leader should move forward, make a quick estimate, and indicate the direction the fire team should continue to clear. This is typically toward the majority of the fortification or toward suspected command and control emplacements. A marker should be emplaced (typically specified in the unit SOP) pointing toward the direction of the cleared path. After employing a grenade, the number two Marine begins moving in the direction indicated and assumes the duties of the number one Marine. Any time the number one Marine runs out of ammunition, they shout, *Reloading!* and immediately move against the wall of the trench to allow the number two Marine to take up the fire. Squad leaders continue to push uncommitted fire teams forward, securing bypassed trenches and rotating fresh teams to the front. It is important to note that trenches must be cleared in sequence vice simultaneously to reduce the chance of fratricide and to ensure no enemy positions are bypassed.

Fire teams use variations of the combat formations described in chapter 6 to move. These formations are used when clearing buildings as well. For these purposes, the terms hallway and trench may be used interchangeably.

Consolidation. As part of consolidation, the leaders must follow the same considerations as for an attack. They must assume a defensive posture to repel anticipated counterattack and conduct site exploitation and a systematic search of secured positions for booby traps and tunnels. This information will be helpful for the HHQ intelligence officer, or for the unit if it occupies the position for an extended length of time.

CHAPTER 4
DEFENSIVE OPERATIONS

PURPOSE OF THE DEFENSE

A defensive operation is conducted to defeat an enemy attack, gain time, economize forces, and develop conditions favorable to offensive or stability actions. Defensive operations alone cannot typically achieve a decision. Their purpose is to create conditions that allow friendly forces to regain the initiative and return to the offense. They do so by attriting or fixing the enemy), retaining terrain that is decisive to mission accomplishment, denying a vital area to the enemy, or increasing an adversary's vulnerability as they concentrate mass to attack.

CHARACTERISTICS OF DEFENSIVE OPERATIONS

Successful defensive activities share combinations of the following characteristics:

- Maneuver.
- Preparation.
- Mass and concentration.
- Flexibility.
- Use of terrain.
- Mutual support.
- Defense in depth.
- Surprise.
- Knowledge of the enemy.
- Local security.

These characteristics are also planning fundamentals for the squad when planning for the defense. For more information on the characteristics of the defense, see MCDP 1-0.

Types of Defensive Operations

There are three types of defensive operations with which the squad can participate: the area defense, the mobile defense, and the retrograde. (Additionally, a retrograde can be a delay, a withdrawal, or a retirement.) While the three types of defensive operations are significantly different, they all possess the same characteristics of the defense as discussed in MCDP 1-0. Additionally, all three types employ offensive activities within the defensive scheme of maneuver.

Squad in the Defense

Squads can expect to be assigned to any of the following elements: the front-line squad, the reserve, or the security element (see figure 4-1).

Front-Line Squad. A squad may defend as part of a front-line platoon within the main battle area. In this case, its mission is to stop the enemy by fire forward of the platoon battle position and to repel them by close combat if they reach the platoon battle position.

Reserve. A squad may be assigned as part of the reserve during the defense. As part of the reserve, the squad is typically assigned a fighting position to the rear of the front-line units, supporting them by fire. The fighting position and sector of fire are assigned to concentrate fire in the rear, on the flanks, or into a gap between front-line platoons. As part of the reserve, the squad may also be assigned a fighting position and sector of fire to limit enemy penetrations. As part of the company's reserve platoon, the squad may participate in a counterattack to expel the enemy.

Security Element. The squad may also serve as part of the security element located forward of the platoon battle position. The squad's mission in this capacity is to gain information about the enemy and to deceive, delay, and disrupt their advance.

Occupation of the Defense

The squad occupies defensive positions in accordance with the platoon commander's plan. To ensure an effective and efficient occupation, rifle squads move to their designated locations marked previously by the leaders' reconnaissance. These positions may also appear on the operational graphics. Once in position, each squad leader checks their location on the map to ensure they are complying with the platoon leader's vision. As the squad occupies its position, the squad leader ensures that each fire team occupies their assigned positions. If there are discrepancies between the actual positioning of the fire teams and the plan, corrections should be made immediately.

Once each fire team has occupied its position, the squad leader must inspect the positions to ensure weapons orientation, positioning of the fire teams, and that the defensive plan is thoroughly understood by all Marines in the squad. Squad leaders should not rely solely on updates from subordinates; they should continuously inspect their defensive sectors. For the purpose of command and control, squad leaders must know the location of the platoon commander and the platoon sergeant.

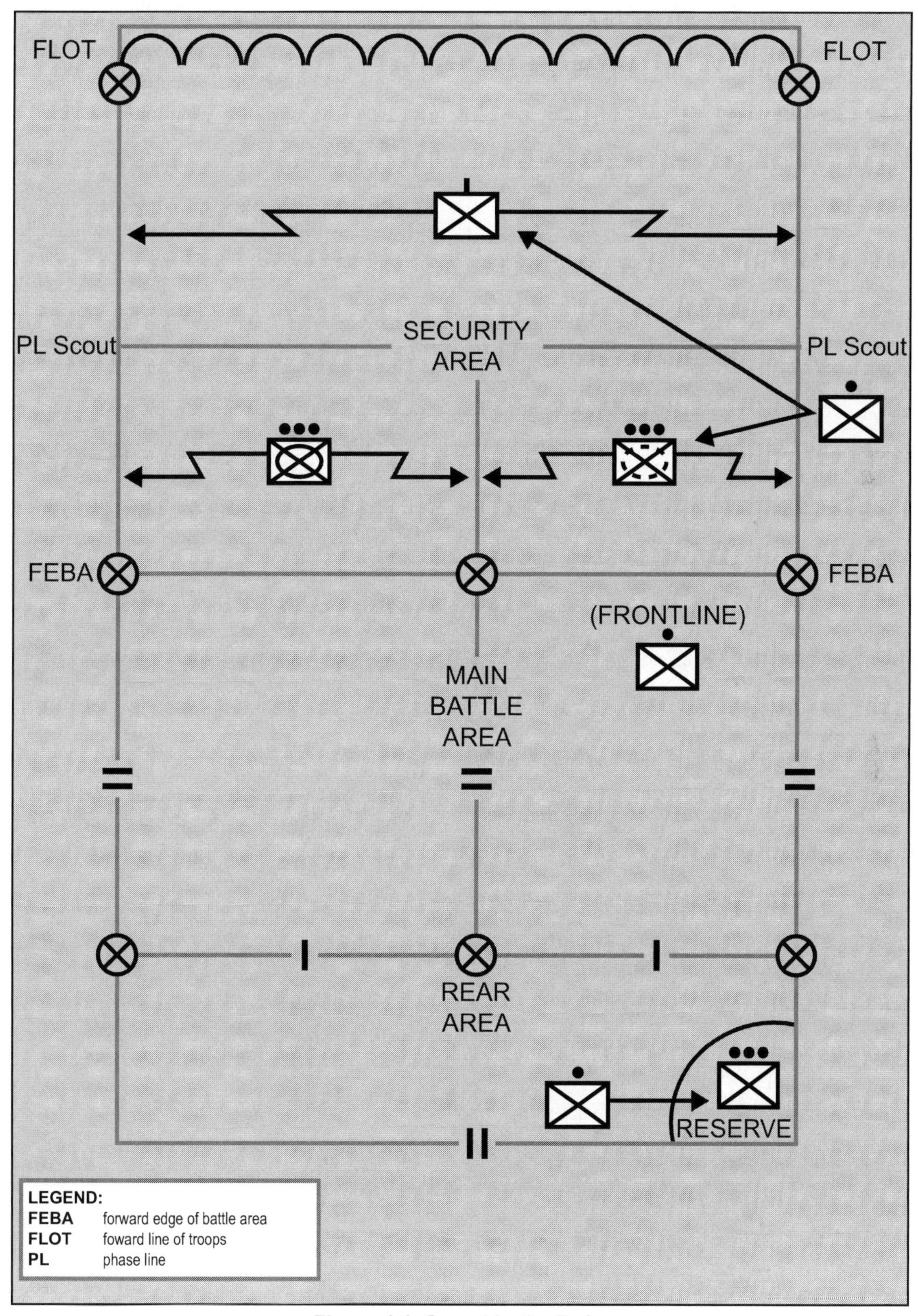

Figure 4-1. Squad in the Defense.

Night vision equipment may aid in occupation under limited visibility conditions. For example, a squad leader can mark their position with an infrared light source, and fire team leaders can move to predetermined positions illuminated by infrared light sources designating their assigned positions. Additionally, squad leaders can use day/night laser pointers (either hand-held or attached to their individual weapons) to point out sectors of fire and target reference points to their Marines to keep the occupation clandestine.

> ***Note:*** Marines must remember that threat forces could possess the same level of technology as friendly forces to detect the use of infrared or thermal devices.

The squad usually occupies the defense in either a hasty or deliberate manner; this is usually based on the time available and the tactical situation.

Hasty. A hastily executed defense is typically organized while in contact with the enemy, or when contact is imminent and the time available for organization is limited. Defensive positions are typically assumed directly from the current positions. The squad may expect to launch hasty attacks to seize favorable terrain on which to establish defensive positions. The hasty defense usually includes—

- Battle positions.
- Hastily established checkpoints (when required).
- Cordons.
- Overnight perimeters.

Hastily constructed defensive positions are continuously improved as the tactical situation permits. The hasty defense should—time and situation permitting—eventually become a deliberate defense.

Deliberate. A deliberate defense is typically organized when out of contact with the enemy, or when contact with the enemy is not imminent and sufficient time for organization is available. A deliberate defense usually includes—

- Fortifications.
- Forward operating bases.
- Permanent checkpoints (i.e., personnel and vehicle).
- Extensive use of obstacles.
- Fully integrated effects of fires (i.e., both lethal and nonlethal).

Once the defensive positions are established, subordinate leaders can begin to develop their sector sketches and fire plans based on the basic fire plan developed during the leader's reconnaissance. Fighting positions should be aggressively improved after approval of the fire plan by the company commander (or platoon commander, if the platoon is operating independently), and upon final inspection of the platoon and squad positions. The platoon commander designates each position's level of preparation based on the time available and other tactical considerations.

In addition to establishing the squad's primary defensive positions, the squad leader and fire team leaders will prepare and plan the occupation of alternate, supplementary, and subsequent defensive positions.

Priorities of Work

Leaders must ensure that Marines prepare the defense quickly and efficiently. Work must be done in priority order to accomplish the most work in the least amount of time while maintaining all-around security and the ability to respond to any enemy action. They priority of work is usually—security, automatic weapons, fields of fire, entrenchment, supplementary and alternate positions, obstacles, and camouflage (see figure 4-2). The following are basic considerations for the priority of work in establishing the defense:

S – Security.

A – Automatic weapons.

E – Entrenchment.

S – Supplementary and Alternative.

O – Obstacles.

C – Camouflage.

Figure 4-2. Priorities of Work in the Defense.

- Establish and emplace local security.
- Position and assign sectors of fire for automatic weapons, grenadiers, and individual riflemen.
- Clear fields of fire and prepare range cards.
- Designate and begin preparing fighting positions.
- Plan for supplemental positions.
- Construct and integrate obstacles.
- Camouflage and conceal positions.

The priority of work is typically found in unit SOPs. However, the company commander may modify the priority of work based on METT-T factors. Several actions may be accomplished at the same time. Squad and fire team leaders must constantly supervise the preparation of the squad's defensive positions.

Security. Security in the defense includes all active and passive measures taken to avoid detection by the enemy, deceive the enemy, and deny enemy reconnaissance elements accurate information on friendly positions. The squad may be tasked with establishing OPs and LPs, conducting security patrols, or a combination of both.

Observation posts are typically located in front of the platoon or squad defensive position along likely avenues of approach to provide early warning and accurate reporting of any enemy activities. An OP is usually within effective small arms range of the platoon's or squad's position (i.e., 300 to 500 meters) depending on the terrain, and is typically manned by no less than two Marines. To avoid possible fratricide, a no-fire area is established around all OPs.

Listening posts are typically established in front of the platoon's or squad's defensive position in the vicinity of the most likely avenue of approach, usually during periods of limited or reduced visibility. As with OPs, LPs are positioned 300 to 500 meters to the front of the platoon or squad, depending on the terrain and visibility. The LPs should be manned by no fewer than two Marines.

The primary purpose of LPs is to provide early warning of enemy approach during periods of limited visibility.

Automatic Weapons Emplacement. To position weapons effectively, squad leaders must know the characteristics, capabilities, and limitations of all company weapons; the effects of terrain; and the tactics used by the enemy. Additionally, the squad leader should consider whether the primary threat will be mounted or dismounted; the plan should address both. Also, the squad leader may have a combination of direct and indirect fire assets from the company attached.

Missiles and Rockets. The primary role of missiles and rockets is to destroy enemy armored vehicles. When there is no armored vehicle enemy threat, they can be employed in a secondary role of providing fire support against point targets, such as crew-served weapons positions.

Infantry Automatic Rifles. The infantry automatic rifle is the squad's primary automatic weapon. They should be laid on the most likely avenues of approach for dismounted forces within the squad's sector of fire and positioned to be capable of delivering effective final protective fires.

Medium Machine Guns. Medium machine guns are the primary crew-served weapon systems that may be attached to a squad. Once attached, the squad leader's first concern is to position the squad around the gun positions to provide the best protection possible. The guns are positioned to place direct fire on locations where the platoon commander wants to concentrate combat power to destroy the enemy. As with the automatic rifles, they should be positioned to deliver final protective fires.

Grenadier Employment. The grenadier is the squad leader's organic indirect fire weapon. The squad leader should position the grenadier to cover dead space in the squad's sector, especially the dead space that cannot be covered by the automatic riflemen or medium machine guns. The grenadier is also assigned a sector of fire overlapping the riflemen's sectors of fire.

Rifleman Employment. Typically, squad leaders should position riflemen to support and protect machine guns and antiarmor weapons. Riflemen are also positioned to cover obstacles, provide security, cover gaps between platoons and companies, or provide observation.

Fields of Fire. In clearing fields of fire forward of the squad's positions, the following guidelines should be considered:

- The squad's position should not be disclosed by excessive or careless clearing; always leave a thin natural screen of foliage to conceal fighting positions.
- Clearing should start just forward of the fighting positions and work forward to the limit of effective small arms fire range.
- In sparsely wooded areas, the lower branches of large trees should be removed. In some situations, it may be desirable to remove whole trees which might be utilized as aiming/reference points for enemy indirect fires.
- In heavy woods, work should be restricted to thinning undergrowth and removing lower branches of large trees. In addition, narrow lanes of fire should be cleared for automatic weapons.

- If practical, buildings and walls forward of the fighting position which may obstruct fields of fire or provide cover and concealment to the enemy should be demolished.
- Care must be taken to ensure that fields of fire are cleared of obstructions which might cause premature detonation of grenadiers' projectiles.

Entrenchment (Construction of Fighting Positions). The defensive plan will typically require the construction of fighting positions. Fighting positions protect Marines by providing cover from direct and indirect fires and by providing concealment from ground and aerial observation through positioning, proper camouflage, and employing deception measures. Because battlefield conditions are never standard, there is no single standard fighting position design that fits every tactical situation. For more information on this and other elements of fieldcraft, see appendix F.

Fighting positions should be constructed even when there is little or no time before contact with the enemy is expected. They should be located behind whatever cover is available and where they can effectively engage the enemy. Positions should provide frontal protection from direct fire while allowing the ability to deliver fire to the front and oblique.

Occupying a position quickly does not mean there is no digging; Marines can dig skirmishers trenches in only a few minutes (see figure 4-3). A fighting position just 18 inches deep can provide a significant amount of protection from direct fire and even fragmentation. The skirmishers trench can be improved over time to a more elaborate and protective position. The squad may typically construct one or a combination of primary, alternate, or supplementary positions. Movement to alternate or supplementary positions should be by covered and concealed routes when possible.

Figure 4-3. Skirmishers Trench.

Primary. Primary fighting positions are—

- The best positions from which to accomplish the mission and cover the assigned sector of fire.
- Assigned to individuals, fire teams, squads, and crew-served weapons.

Alternate. Alternate fighting positions are—

- Not typically assigned to individuals within the squad.
- Utilized primarily by crew-served weapons.
- Positioned so that the squad or crew-served weapons may accomplish their original mission if the primary position becomes unsuitable or untenable.

Supplementary. Supplementary fighting positions are—

- Secondary positions that do not cover the same sector of fire as the primary positions.
- Prepared to guard against attacks from directions other than where the main attack is expected.
- Designed to provide security.

If not prescribed by higher authority, the squad leader designates either individual or two-person fighting positions. The type of fighting position used is usually based upon—

- Squad strength.
- Fields of fire.
- The size of the squad's sector of fire.
- Morale.

Individual Fighting Position. Positions occupied by a single Marine are least desirable, but useful in some situations, such as providing security to crew-served weapons. These fighting positions may have to cover extremely wide frontages, and should never be out of direct sight from adjacent positions to allow for fire to the front or to the oblique from behind frontal cover (see figure 4-4).

Some considerations when using individual fighting positions include—

- Individual positions allow for more choices in the use of cover.
- An individual position only needs to be big enough for one Marine and their gear.
- They take less time to construct.
- They do not allow the level of security afforded by two-person positions.

Two-Person Fighting Position. A two-person fighting position (see figure 4-5) is generally more effective than an individual fighting position. It can be used to provide mutual support to adjacent positions on both flanks and to cover dead space immediately in front of the position. One or both ends of the hole may extend around the sides of frontal cover.

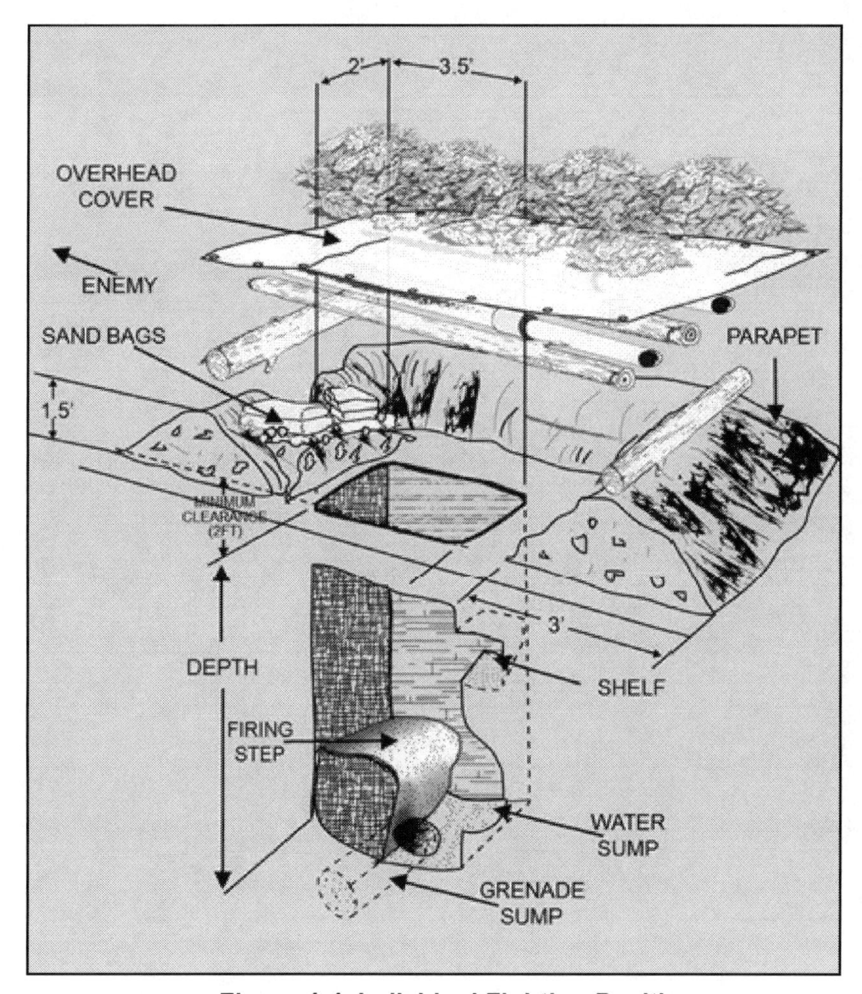

Figure 4-4. Individual Fighting Position.

Adapting a position in this way allows both occupants to have better observation and greater fields of fire to the front. Also, during rest or other required activities, one Marine can watch the entire sector while the other sleeps or performs other tasks. If they receive accurate fire from the front, they can move back to take advantage of the protection of the frontal cover.

The two-person fighting position—

- Requires more time and materials to construct.
- Is more difficult to camouflage.
- Provides a larger target for enemy forces.

Considerations. The most important consideration of a fighting position is that it must be tactically positioned. Squad and fire team leaders must be able to look at the terrain and quickly identify the best location for fighting positions. Good positions should—

- Provide the ability to engage the enemy within the Marines' assigned sectors of fire.
- Provide the ability to fire out to the maximum effective range of individual weapons with maximum grazing fire and minimal dead space.
- Allow grenadiers to be placed in positions to cover dead space.

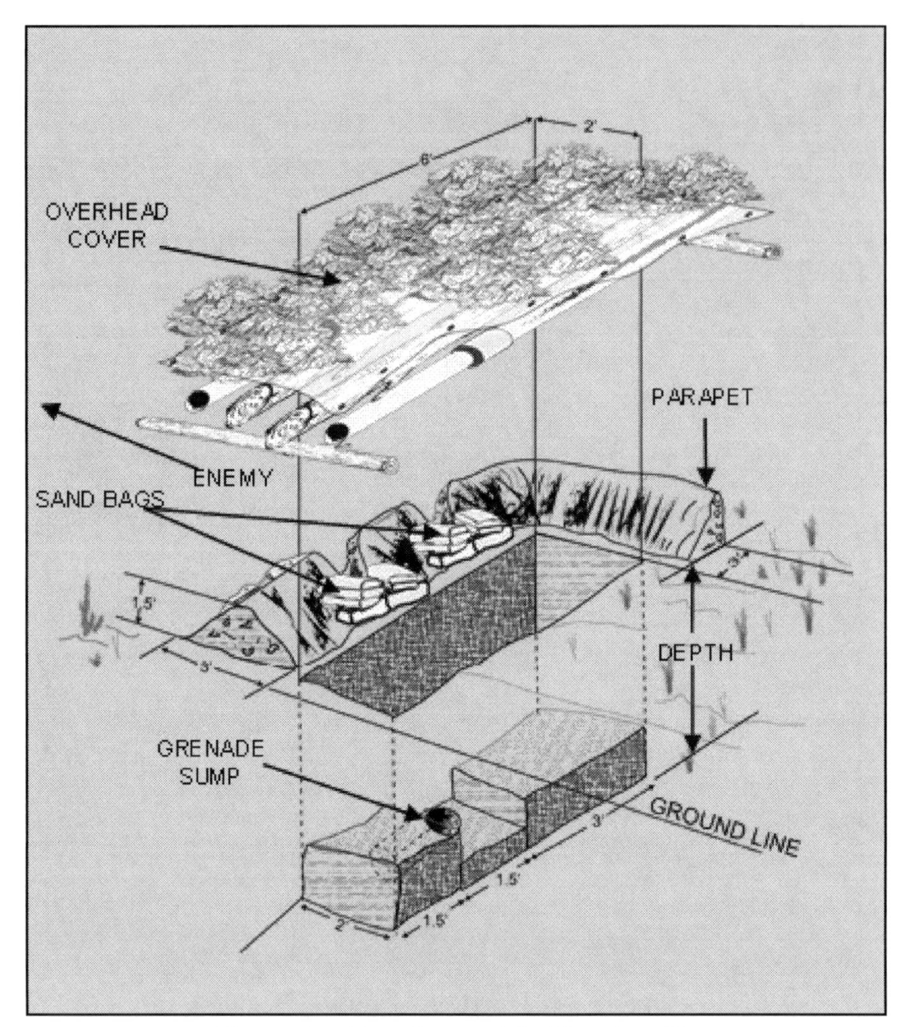

Figure 4-5. Two-Person Fighting Position.

Fighting positions should be mutually supportive and provide interlocking fires with other adjacent squads and units. If the squad is operating independently, its position should provide all-around security. When possible, fighting positions should be sited behind available natural cover and in easily camouflaged locations. The enemy must not be able to identify the position until it is too late and they have been effectively engaged.

Preparation of Fighting Positions

Squad and fire team leaders must ensure their Marines understand when and how to prepare fighting positions based on the immediate tactical situation. Fighting positions should be constructed every time the squad conducts an extended halt—half the squad digs in while the other half maintains security. Marines should prepare fighting positions in steps, with leaders inspecting progress at each step before Marines move to the next.

When anticipating immediate enemy contact, the squad should dig skirmishers trenches. As time is available, these defensive positions can be continually improved, enlarged, and strengthened.

Step 1. Unless directed by the platoon commander, squad leaders should check fields of fire from the prone position prior to assigning sectors of fire. Once sectors of fire are designated, individual Marines should perform the following:

- Emplace sector stakes (e.g., tent poles, tree branches, sandbags) for limited visibility engagement of targets.
- Position aiming stakes within Marines' primary sectors.
- Position depression stops, (e.g., logs, sandbags) for grazing fire between sector stakes.
- Trace the outline of the fighting position on the ground (approximately seven feet wide for a two-person fighting position).
- Clear fields of fire for the assigned sector.

Step 2. Marines should prepare retaining walls to support the construction of parapets to the front, flanks, and rear of their fighting positions. They should ensure the following:

- There is a minimum distance (approximately the width of one helmet) between the edge of the hole and the beginning of the front, flank, and rear parapets.
- The cover to the front consists of sandbags (or logs), two to three high. For a two-person fighting position, the cover should be about the length of two service rifles (about seven feet).
- The cover to the flanks is the same height, but only one service rifle length (about 3.5 feet).
- The cover to the rear is one sandbag high and one service rifle long (about 3.5 feet).
- If logs are used, they must be held firmly in place with strong stakes.

Step 3. Marines begin digging the fighting position, throwing the soil forward of the parapet retaining walls, and packing it down (see figure 4-6). The following should be accomplished:

- Dig the position should be dug armpit deep (for the tallest Marine) to the firing step.
- If using turf or topsoil to camouflage the parapet, remove sufficient ground cover and set aside.
- Fill parapets in order: front, flanks, and rear.

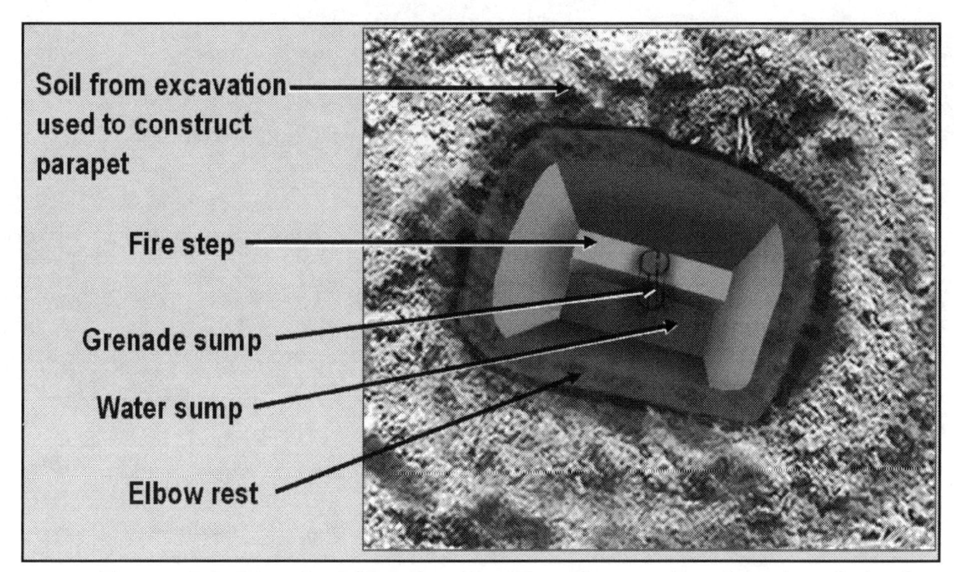

Figure 4-6. Step 3 of Fighting Position Preparation (overhead view).

- Camouflage the parapets and the entire position.
- Dig grenade and water sumps and slope the floor toward them.
- Dig shelves for hand grenades and storage areas for two rucksacks into the rear wall as needed

Grenade Sump. A grenade sump should be cone-shaped, with the opening measuring approximately as wide as the spade of the entrenching tool, narrowing to about five inches in diameter at the end. It should slope downward at an angle of 30 degrees, and should be as deep as the Marine can make it.

Water Sump. A water sump is dug at one side of the fighting hole below the firing step to collect water and provide a space for the Marines' feet while seated on the firing step.

Step 4. Marines begin preparing overhead cover and concealment (see figure 4-7). Due to the proliferation and increased capability of small UASs to conduct reconnaissance, surveillance, and target acquisition missions, squad leaders must construct overhead cover and concealment, combined with other deception measures to conceal the squad's fighting positions from aerial observation and fires. Terrain often allows the construction of positions with overhead cover and concealment that protects Marines from indirect fire, fragmentation, and observation, while still allowing the ability to return fire. In some occasions, particularly on open terrain, this may not be possible, and the entire position must be built below ground level. Although this type of position offers excellent protection and concealment, it limits their ability to return fire from within a protected area. To prepare overhead cover and concealment, Marines should—

- Always provide solid lateral support. Build the support with 4- to 6-inch logs (or equivalent materials) on top of each other running the full length of the front and rear cover.
- Place five or six logs that are four to six inches in diameter and two service rifles long (approximately seven feet) over the center of the position, resting on the overhead cover support (not on the sandbags).

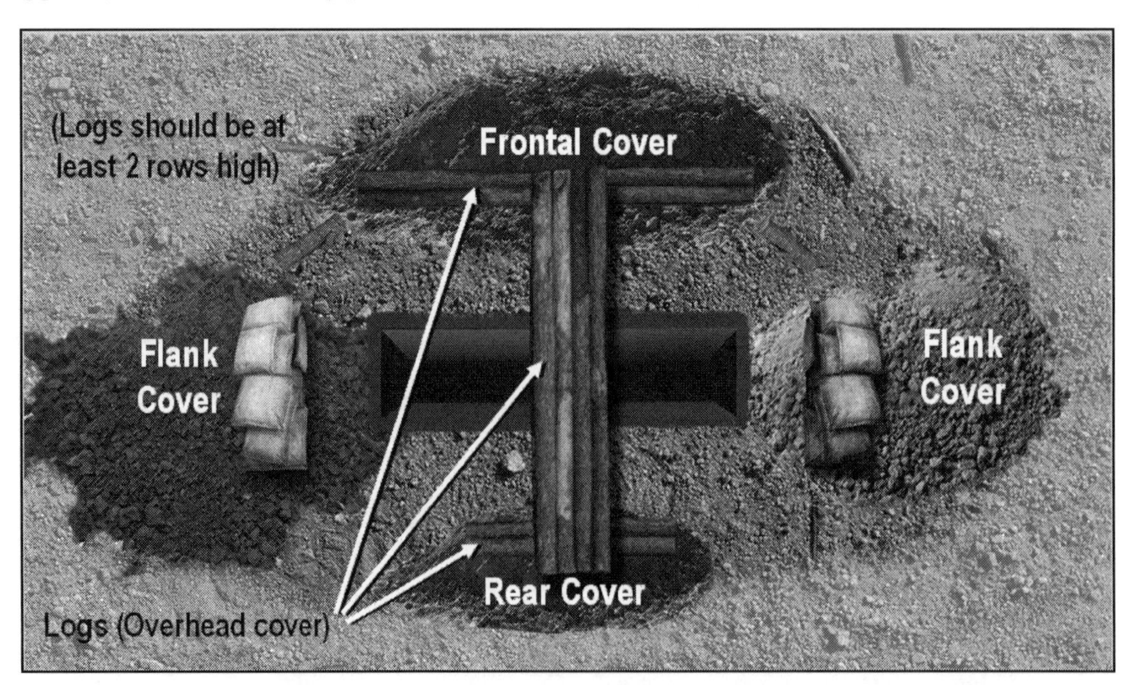

Figure 4-7. Step 4 of Fighting Position Preparation: Construction of Overhead Cover.

- Place waterproofing material (e.g. plastic bags, ponchos) on top of these logs.
- Place a minimum of 18 inches of packed dirt or sandbags on top of the logs.
- Camouflage the overhead cover and the base of the position with natural or artificial materials.
- Leaders must inspect the positions.

Selecting Supplementary and Alternate Fighting Positions. The squad prepares supplementary fighting positions that are organized in the same manner as the primary fighting positions but oriented in a different direction. If crew-served weapons are attached or employed in the squad's sector, alternate fighting positions should also be prepared for them.

SQUAD FIRE PLAN

The squad leader formulates the squad fire plan so as to physically occupy the squad's assigned primary fighting position, allowing the squad the ability to cover the sector of fire assigned by the platoon commander. The fire plan includes the following at a minimum:

- Assignment of fire team sectors of fire.
- Fire team fighting positions.
- Principal directions of fire for the automatic rifles.
- The squad leader's fighting position.

The squad leader assigns a principal direction of fire for each automatic rifle if one is not assigned by the platoon commander. The squad leader selects the exact fighting position for each automatic rifle. Squad leaders usually occupy positions slightly to the rear of the fire teams from where they can best control their squads' positions. The squad leader's position should be a position from which the squad leader can—

- Observe the squad's assigned sector of fire.
- Observe as much of the squad fighting position as possible, particularly the positions of the fire team leaders.
- Maintain contact with the platoon commander.

Fire Plan Sketch

Squad leaders prepare the squad fire plan sketch in duplicate, giving one sketch to the platoon commander for approval and keeping a copy for themselves which they maintain at their fighting positions. The sketch should include the fire team fighting positions and sectors of fire, the fighting positions and principal directions of fire of the automatic rifles, and the squad leader's fighting position. If the rifle squad is providing protection for a crew-served weapon, its position, primary fire mission (i.e., final protective line for machine guns and principal direction of fire for other crew-served weapons), and range card should be included as part of the sketch. Figure 4-8 is an example of a squad fire plan sketch.

Range Cards

A range card is a rough sketch or drawing which serves as both a record of firing data and a document for defensive fire planning (see figure 4-9). Gunners use the card to recall data required to deliver fire upon predetermined targets, and as an aide in estimating ranges to other targets within their sector of fire. Like the squad fire plan sketch, the range card is made in duplicate. Regardless of the amount of time gun crews expect to occupy a position for, a range card should be prepared immediately. For more information on range cards, see MCTP 3-01C, *Machine Guns and Machine Gunnery.*

The range card should be drawn to scale and contain the following:

- Orientation to the ground.
- The gun number, unit designation (usually no higher than company), and date prepared.
- Either a primary or alternate position.
- Final protective fires (i.e., either a final protective line or a principal direction of fire, whichever is assigned).

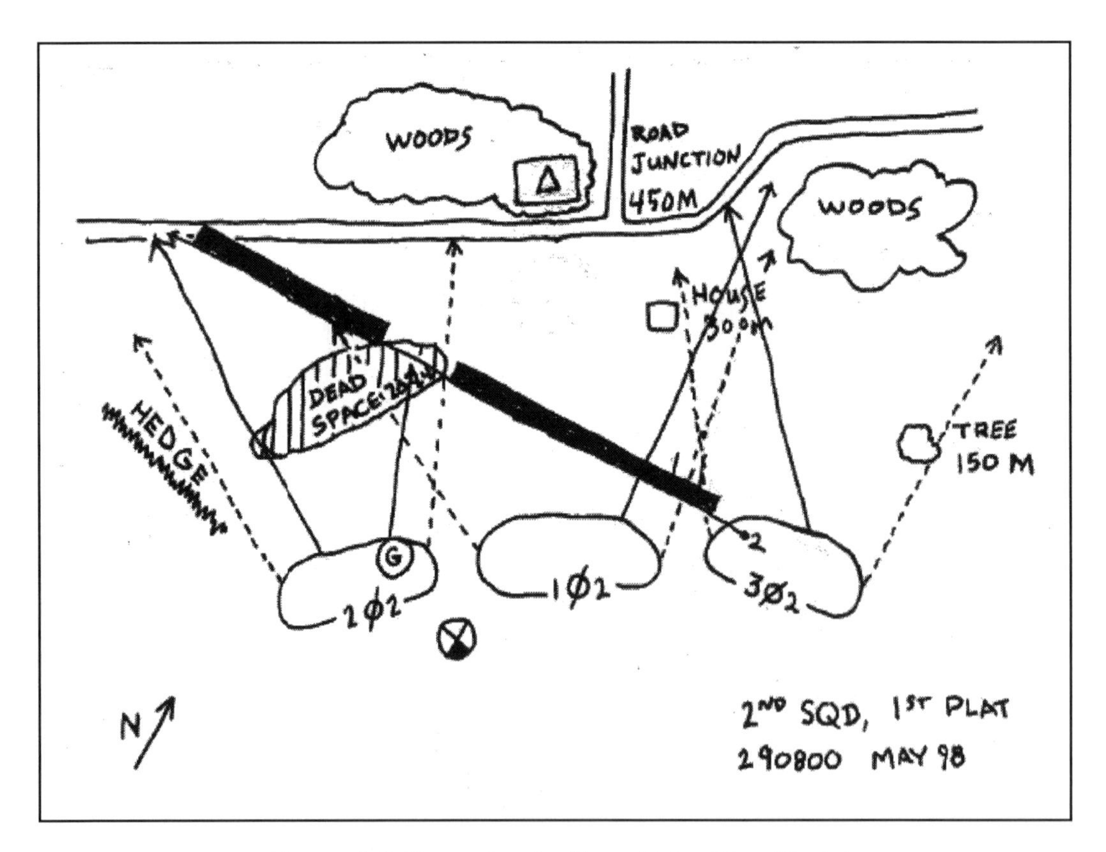

Figure 4-8. Example of a Squad Fire Plan Sketch.

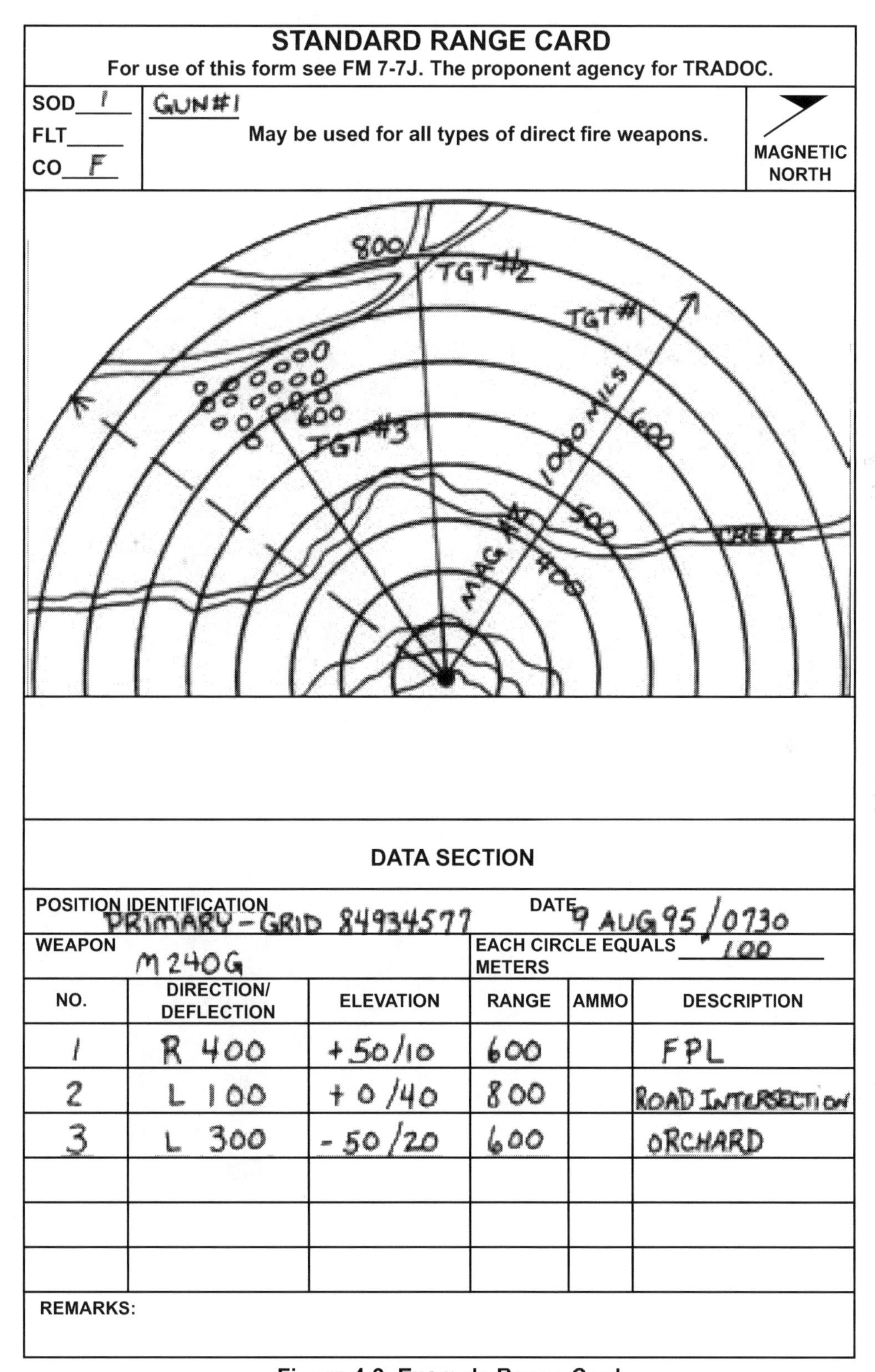

STANDARD RANGE CARD

For use of this form see FM 7-7J. The proponent agency for TRADOC.

SOD 1 | GUN #1 | May be used for all types of direct fire weapons. | MAGNETIC NORTH

FLT

CO F

DATA SECTION

POSITION IDENTIFICATION PRIMARY - GRID 84934577

DATE 9 AUG 95 / 0730

WEAPON M240G

EACH CIRCLE EQUALS 100 METERS

NO.	DIRECTION/ DEFLECTION	ELEVATION	RANGE	AMMO	DESCRIPTION
1	R 400	+50/10	600		FPL
2	L 100	+0/40	800		ROAD INTERSECTION
3	L 300	-50/20	600		ORCHARD

REMARKS:

Figure 4-9. Example Range Card.

CHAPTER 5
PATROLLING

PURPOSE OF PATROLLING

A patrol is a detachment sent out by a larger unit to gain information about the terrain and enemy, provide security for defensive positions, or cause attrition of enemy forces and materials by ambush. Patrols operate semi-independently, returning to the main body upon completing their mission. Patrolling facilitates locating the enemy to either engage them or report their disposition, location, and actions.

A commander must have current information about the enemy and terrain in order to employ their unit effectively. Patrols are an important means of gaining this information, and are used to destroy enemy installations, capture enemy personnel, perform security missions, or prevent the enemy from gaining information.

Modern warfare places a high premium on effective patrolling because units have larger areas of operation and can be threatened from all directions. As the distance between units increases, aggressive patrolling becomes necessary for security, to prevent infiltration by enemy units, and to establish contact with friendly adjacent units.

The mission to conduct a patrol may be given to a fire team, squad, or platoon. The keys to successful patrolling include—

- *Detailed planning*. Every portion of the patrol must be planned and all possible contingencies considered.
- *Productive, realistic rehearsals*. Each phase of the patrol is rehearsed, beginning with actions in the objective area. Similar terrain and environmental conditions are used when conducting rehearsals.
- *Thorough reconnaissance*. Ideally, the patrol leader will physically conduct a reconnaissance of the route and objective. Photographs and/or maps will be used to supplement the reconnaissance.
- *Positive control*. The patrol leader must maintain positive control. This includes supervision during patrol preparations.
- *All-around security*. Security must be maintained at all times, particularly near the end of the patrol where there is a natural tendency to relax.

PATROL ORGANIZATION

The platoon commander designates a patrol leader, typically one of the squad leaders, and gives them a mission. The patrol leader then establishes the patrol headquarters and the patrol units required to accomplish the mission. These major subdivisions are called elements. Personnel are assigned to units based on the mission of the patrol and the mission of individuals within the patrol. The existing infantry structure is reinforced as required. More specific information about patrol organization is contained below. The major elements include—

- *Patrol headquarters*. The patrol headquarters is composed of the patrol leader and personnel who provide support for the entire patrol, such as the—
  - Corpsman.
  - Communicator or radio operator.
  - Forward observer.
- *Security element*. In a large patrol, the security element can provide all-around security to the front, flanks, and rear. To accomplish this distributed task, it must be provided with an effective means of communication or arrange appropriate visual signals through the nearest element of the patrol. The alternative is that the security element provides point security, the assault element (middle) provides flank security and the support element (or trail) provides rear security. For smaller patrols such as squad-sized, security en route to the objective is the responsibility of every individual in the patrol. At the objective area, the security element provides security for the objective rally point (ORP) and to the flanks and external avenues of approach to the objective.
- *Support element*. The support element provides support by fire for the attack, covering fire for the withdrawal, and supporting fires to cover danger area crossings throughout the patrol.
- *Assault element*. The assault element attacks and seizes the objective. It also provides searchers to clear the objective, pacers, a compass Marine, a navigator, and an assistant patrol leader en route to and back from the objective area.
- *Reconnaissance and surveillance element*. The mission of this element is to gather information to answer requests for information/intelligence to support commanders' decision making and mission accomplishment. This element is METT-T dependent.

Any attachments a patrol may have are added to the element that best supports its function. Attachments may include, but are not limited to, the following:

- Demolition teams.
- Scout snipers.
- Machine gun squads/teams.
- Translators.
- Military working dogs with handlers.

Task Organization

The squad leader must plan for maximum flexibility to handle any emergency and ensure that the patrol's mission is not put in jeopardy with the loss of key members, a team, or an entire element. Depending on the METT-T analysis, elements in the organization can be further subdivided into teams, each of which performs essential designated tasks. In creating teams, unit integrity should be maintained. The patrol is organized so each individual, team, and element is assigned a specific task, but capable and prepared to perform other tasks. This may not be possible for certain special tasks requiring a trained specialist or technician, such as electronic warfare.

A squad-sized or reinforced squad-sized patrol can move in any of the normal squad formations, but would typically move in squad column (see figure 1-15). The order of march is typically security, then assault, then support, but may be modified as the squad/patrol leader sees fit. Also, the patrol leader may modify the formation to allow for point, flank, and rear security. Additionally, a patrol may find it useful to utilize a single-file or "ranger" file for better control if the terrain dictates and the tactical situations permits it.

Navigation Team. Usually located within the lead element, the navigation team is responsible for the overall navigation of the patrol. Personnel in this team must be proficient in land navigation and have a thorough knowledge of all checkpoints, tactical control measures, and routes to be utilized during the patrol and be able to communicate the current location to the patrol leader as required. The team should carry special equipment to assist them in the execution of these duties (e.g., maps, lensatic compasses, and protractors).

Aid and Litter Team. Usually located within the support element, the aid and litter team is responsible for rendering aid to and evacuating wounded personnel during the conduct of the patrol. Marines must have a thorough knowledge of the casualty care and evacuation plan as outlined in the patrol order. The team should carry special equipment to assist them in the execution of these duties (e.g., combat lifesaver bags, stretchers, and pole-less litters). Marines must be prepared to establish snap casualty collection points, assist the corpsman in caring for casualties, and be knowledgeable about CASEVAC 9-line reports.

Detainee Handling Team. The detainee handling team is responsible for searching, transporting, and safeguarding captured enemy personnel. Personnel must have a thorough understanding of the 5 S and a T (i.e., search, segregate, silence, speed, safeguard, and tag) of detained handling, as well as the detainee handling plan outlined in the patrol order. The team should carry special equipment to assist them in the execution of these duties (e.g., flexi-cuffs, cameras, evidence bags, and tags for the documentation of captured materials).

Demolition/Breach Teams. This team plans and executes the destruction of obstacles and enemy equipment and creates small scale breaches in enemy protective obstacles to facilitate the completion of the squad's primary task.

Landing Zone Team. The landing zone (LZ) team is responsible for the establishment of LZs during the conduct of the patrol. Personnel must have a thorough knowledge of the various methods and techniques of establishing and marking such zones, particularly as outlined in the patrol order (see appendix C). The team should carry special equipment to assist them in the execution of these duties (e.g., infrared "buzz saw" and marking panels).

> *Note:* It is important to note that an element may be tasked with more than one special task (e.g., the aid and litter team may double as the LZ team). The squad leader must take care when organizing the patrol to structure and task the unit to best support the mission.

Duties of Personnel

In addition to organizing the elements and special teams, the squad leader further organizes the patrol by tasking individual members with various roles and responsibilities. In order to maintain unit integrity, care must be taken to ensure that the member tasked is assigned to the team that their responsibilities best support.

Patrol Leader. Designated by the platoon commander, a patrol leader is responsible for the execution of the patrol from the time the initial mission is received until the patrol has reentered friendly lines and been properly debriefed. The patrol leader is responsible for organizing, planning, preparing for, and conducting the patrol throughout the duration of the mission.

Assistant Patrol Leader. Designated by the patrol leader, the assistant patrol leader may be an independent billet or may be assigned as an additional duty to a fire team leader. The assistant patrol leader supports the patrol leader as required and is often the driving force in preparing the patrol during the time between the publication of the warning order and the issue of the patrol order. They serve as a force multiplier, freeing up the patrol leader to be a tactical leader by covering the administrative and logistical actions required of the mission. During execution, the assistant patrol leader assists the patrol leader in maintaining control and accountability. Due to the importance of the assistant patrol leader billet, it is important that the patrol leader takes great care when selecting them. The assistant patrol leader must be ready to instantly take control of the patrol should the patrol leader become a casualty, and thus must be thoroughly knowledgeable in all aspects of the patrol leader's plan.

Fire Team Leaders. During preparation, the fire team leaders assist the patrol leader by overseeing the preparation of their fire teams. During execution, the fire team leaders lead their teams as tasked by the patrol leader. Fire team leaders must be ready to assume the duties of the assistant patrol leader if the need arises, and thus must thoroughly understand all aspects of the mission.

Navigator. The navigator, generally located within the lead element as part of the navigation team, is responsible for keeping the patrol on the assigned route throughout the duration of the patrol. At a minimum, the navigator should carry a map, protractor, and lensatic compass, even if equipped with a GPS [global positioning system]. The navigator must be able to readily identify checkpoints and rally points throughout the assigned route, as well as be able to give the patrol leader an accurate position report at any given moment during the patrol. Because of this, the navigator must have a thorough understanding of all checkpoints, tactical control measures, and routes to be utilized during the patrol. An assistant navigator must be assigned to assist the navigator in the execution of these duties.

Pacer. Generally located with the lead element as part of the navigation team, the pacer works closely with the navigator and is responsible for keeping track of the distance the patrol has traveled. The pacer must be familiar with the assigned routes and have a thorough knowledge of the distance between checkpoints. The pacer must maintain a current and accurate pace count over a wide range of terrain in a wide range of conditions in order to accurately measure the distance the patrol has traveled. The pacer may serve additionally as the assistant navigator.

The Point. The point is the forward most individual in the patrol's formation, regardless of the order of movement. The position is not utilized as a trail breaker. The point is responsible for guiding the patrol along the assigned route, being careful to identify any obstacles or threats in the patrol's path. If an obstacle is easily bypassed, the point leads the patrol around it. In many instances, the point is the first patrol member to encounter a situation that directly effects the unit's execution of the patrol. The point must be able to rapidly assess the situation and determine the immediate course of action. Because of the importance of this billet, patrol leaders must ensure that they have vetted potential points' abilities well in advance to ensure they are suited for the task. The patrol leader may choose to assign an assistant point, who may be tasked with providing local security for the point during searches of booby traps and other obstacles.

Flank Security. If the terrain or situation warrants, the patrol leader may choose to assign one or more individuals as flank security. These members travel parallel to the main body of the patrol at a distance that allows the patrol leader to maintain command and control. The distance may vary according to terrain and visibility. The flank security acts as an early warning measure to the main body and prevents the enemy from ambushing or infiltrating the patrol from the flanks.

Communicator. If a communicator is not available, the patrol leader may assign a Marine within the patrol as the communicator. The communicator is responsible for establishing and maintaining communication with other units, elements, and agencies as designated in the patrol order. The communicator is responsible for transmitting appropriate reports and relaying any relevant transmissions received to the patrol leader. The communicator must be ready at any time to call for CASEVAC. The communicator, who serves as the patrol's lifeline with all other agencies, should be positioned in the patrol leader's immediate vicinity. The communicator must be thoroughly knowledgeable in all reports and reporting procedures.

Patrol Leader's Preparation

During planning, the squad leader uses the troop leading procedures, aided by checklists and the unit SOPs, to ensure that all required events are planned for and all patrol members know their duties. The normal sequence for the patrol steps is listed below, but the sequence may vary depending on the availability of personnel, the timing of a leader's reconnaissance, and the extent of coordination already made with the assistance of the platoon commander or company commander. The typical sequence for planning and executing a patrol is to—

- Study the mission.
- Plan use of time.
- Study the terrain and situation.
- Organize the patrol.
- Select personnel, weapons, and equipment.
- Issue the warning order.

- Coordinate (continuous throughout).
- Make a reconnaissance.
- Complete detailed plans.
- Issue the patrol order.
- Supervise (continuous), inspect, rehearse, and re-inspect.
- Execute the mission.

Study the Mission. The squad leader carefully studies the mission and all other information provided by the platoon commander. In so doing, the squad leader identifies other significant tasks (i.e., implied tasks) which must be accomplished in order for the squad to accomplish its primary mission. These implied tasks are further identified as assignments for the patrol's elements and teams and may require special preparation, planning, personnel, and/or equipment. For more information on tactical tasks, see appendix G.

Plan Use of Time. In order for the squad leader to properly use the time allotted from receipt of the order until departure from friendly lines, they prepare a schedule which includes every event that must be done prior to departing friendly lines. In preparing the schedule, the squad leader uses reverse planning, working backwards from the time of departure to the present. Some of the key events that must be done are:

- Making a leader's reconnaissance of the patrol routes and the objective (time permitting).
- Pre-combat checks and pre-combat inspections of the patrol.
- Rehearsals (i.e., day and night).
- Issuing the patrol order.

Study the Terrain and Situation. The squad leader makes a thorough map study of the terrain over which the patrol will operate. The terrain in the vicinity of the objective will influence the patrol organization, to include its size, equipment (both standard and special), the manner in which the squad leader will conduct the reconnaissance, and the disposition of the squad at the objective. The squad leader studies the friendly and enemy situations to determine the effects that their dispositions, strengths, and capabilities may have on the mission. These factors also influence the routes, the organization of the squad, and the weapons and equipment to be taken.

Organize the Patrol. Organization consists of determining the units and teams required to accomplish the essential and implied tasks. Organization is a two-step process; the steps include general organization and special organization.

Select Patrol Members. Typically, squad leaders are limited to selecting patrol members from within their own squad, with the addition of those special personnel/teams made available by the platoon commander. If the patrol leader is a squad leader, most likely everyone in the squad will participate in the patrol. The squad leader should maintain fire team integrity whenever possible.

Select Weapons. Patrol members are usually armed with their organic weapons. The squad leader should request the support of specific weapons teams when required for tasks beyond the capability of the squad (e.g., machine gun teams for ambushes).

Select Equipment. There are five general purposes or areas of consideration when the squad leader chooses equipment:

- *In the objective area*. This is the equipment with which the squad accomplishes its mission. It includes such items as ammunition (i.e., the number of rounds per Marine), demolitions (i.e., the type and amount), binoculars, NVDs, cameras, pyrotechnics, and biometric devices.
- *En route*. This is equipment that enables or assists the squad to reach its objective. It includes items such as maps, compasses, binoculars, wire cutters, ropes, flashlights, and ammunition.
- *Control*. This equipment is used in assisting the squad leader and team leaders in controlling the squad while conducting movement to and during actions at the objective area. It includes such items as whistles, pyrotechnics, communications devices, chemical (chem) lights, and luminous tape.
- *Routine equipment*. This is the typical equipment carried by all squad/patrol members, usually based on unit SOP. It includes the uniform to be worn, and individual clothing and equipment to be carried.
- *Water and food*. The patrol leader specifies the number of water bladders and full canteens to be carried by all patrol members. Rations should be issued for all patrols, regardless of duration, as part of contingency planning. Rations are usually "field stripped" of nonessential items to reduce weight and possible battlefield clutter (i.e., target indicators).

Issue the Warning Order. Thus far, the squad leader has been going through the mental processes necessary to arrive at some initial conclusions and develop a mental picture about the mission. The squad leader has determined how the squad will be organized and what attachments will be needed. They have determined a rough time schedule, conducted a thorough map reconnaissance, and identified implied tasks that must be accomplished if the squad's mission is to be successful. Now, the squad leader must inform the squad members so they can begin preparations.

Ideally, the squad leader issues the warning order to all patrol members, including attachments. If that is not possible, the squad leader must ensure that all fire team and attached unit leaders are present to receive the warning order.

Coordinate (Constant and Continuous Throughout). The squad leader begins coordination from the time they receive the order. They may receive assistance from their platoon commanders, platoon sergeant, or other squad leaders. Squad leaders should be primarily concerned with the following:

- *Movement in friendly areas*. The squad leader finds out about the locations and activities of other friendly units or patrols operating in the area so as not to restrict or endanger the patrol during movement; routes and fires are planned accordingly.
- *Departure and re-entry of friendly lines/areas*. The squad leader checks with the small unit leaders occupying the areas through which the patrol will depart and return and whether or not guides from their units will be required to lead the patrol through any friendly obstacles, such as mines or wire.
- *Fire support*. During the briefing with the platoon commander, the squad leader should find out what fire support is available during the patrol (i.e., both lethal and nonlethal effects) and what pre-planned artillery and mortar targets have already been plotted along the routes to and from

the objective area and within the objective area itself. Next, the squad leader plans for additional fires (if necessary) along the patrol's route to the objective area, at the objective area, and to cover the withdrawal from the objective area back to the unit's position. Also, they determine the availability of any UASs, armed or unarmed, that may be operating within the area during the time of the patrol. If applicable, the squad leader must coordinate with the company combat operations center regarding the latest IO messaging being utilized, to include any pamphlets, handbills, or other materials for distribution.

- *Logistical support*. The squad leader must make arrangements for the delivery of any emergency resupply for contingency purposes (e.g., ammunition, special equipment, demolitions, water). Squad leaders must also inquire as to the use of air assault assets for both CASEVAC and emergency extract during the patrol's movement and at the objective area.
- *Information checklist*. The squad leader should try to find out as much information about the enemy and any local inhabitants that may live there as possible. Specifically, the following should be determined:
  - What is the enemy's pattern of operations (e.g., conducting patrols, ambushes)?
  - What type weapons do they have?
  - What is their strength and disposition?
  - What are the enemy's tactics, techniques, and procedures (e.g., do they use mines and booby traps)?
  - Are the locals supportive of the enemy forces?

The squad leader must coordinate with the CLIC to answer as many requests for information as possible. If it's available, the squad leader should contact the battalion intelligence section (S-2). They may be able to provide valuable information about the enemy, including anything obtained from previous patrol reports, aerial photographs, satellite imagery, or reconnaissance operations.

Make Reconnaissance. Whenever possible, the squad leader should make a physical reconnaissance of the routes to follow and of the objective area. However, this is often not possible because of the enemy situation and/or the availability of either air assault, mechanized, or motorized assets. Instead the squad leader must rely on a map reconnaissance and information gathered from other sources.

Complete Detailed Plans. The squad leader is now ready to plan the patrol in detail. Through discussions with the platoon commander and additional coordination, the squad leader has already determined the situation and the mission. The remaining planning deals with how the patrol is to be executed, the tasks assigned to each element/team, administrative and logistical matters, and command and signal. The squad leader needs to apply special attention to the following:

- *Specific duties of elements, teams, and individuals*. The warning order assigned general tasks to elements, teams, and key individuals. The squad leader now assigns specific duties to each of these entities.
- *Primary and alternate routes*. The squad leader selects patrol routes based on map studies, aerial imagery, the leader's reconnaissance (if possible), and coordination with others who have been over the terrain. The squad leader should always plan a primary and alternate route to the

objective. Whenever practical, the patrol leader should plan the return by way of a different route. Patrol routes are pointed out to the patrol members by—

- Indicating the routes on a map or overlay.
- Designating objectives, checkpoints, and rally points.

- *Conduct of the patrol*. The squad leader's plan must address all the following:
  - Patrol formation and order of movement.
  - Departure from and re-entry to friendly lines or areas.
  - Rally points and actions at rally points.
  - Location of the ORP and actions to be taken there.
  - Actions at danger areas.
  - Actions on enemy contact.
  - Actions at the objective.
- *Arms and ammunition*. The squad leader checks to see if the arms and ammunition specified in the warning order have been obtained.
- *Uniform and equipment*. The squad leader checks to see if all required equipment was available and was drawn.
- *Casualties and detainees*. The procedures for handling wounded vary, depending on the seriousness of the wound and if the team is en route to the objective, at the objective, or returning to friendly areas. A patrol might continue to the objective carrying its casualties, send them back with a detail of Marines, return with the casualties, or call for assistance. Personnel who become casualties at the objective area or on the return to friendly areas will typically be transported by whatever means are available—carried by the patrol, in vehicles, or by aircraft. Detainees typically travel with the patrol, guarded by personnel designated in the patrol order.
- *Signals*. The squad leader's plan addresses the type of signals to be used during the patrol—hand-and-arm signals, communication devices, pyrotechnics, and audible. A primary and alternate signal for each event requiring signals should typically be planned for.
- *Communication with higher headquarters*. The squad leader includes all essential communication details—call signs, frequencies, reporting times (usually upon reaching checkpoints), brevity codes, and security requirements.
- *Challenge and password*. The HHQ will designate the challenge and password for the unit.
- *Location of leaders*. The squad leader plans the location from where to best control the patrol, usually in the forward one-third of the formation. The assistant patrol leader is placed where they can best assist in control during movement, usually near the rear of the formation. At the objective, the assistant patrol leader assumes a position from where to readily take command should the squad leader become a casualty.

Issue Patrol Order. After completing the plan, the squad leader assembles the members of the patrol and issues the order, presenting it in a clear, concise manner following the standard five-paragraph order format (see appendix C) and ensuring that—

- All patrol members are present.
- The unit/team leaders provide a status report on the preparatory tasks assigned to them in the warning order.

- An orientation precedes the issuance of the order.
- A terrain model is used when possible to help explain the concept of operations for the movement to the objective area, the actions at the objective area, and the return to friendly lines/areas.
- The entire order is issued before taking questions.
- At the conclusion of the question and answer session, a "time hack" is given and the time of the next event is announced. For example:
  "It is now 1700. Everyone get some chow and I'll inspect the patrol at 1745, in movement formation, in that clump of pines near the company command post."

Every Marine a Collector. The ability to maximize the observation and reporting skills of individual Marines is critical to successful patrol execution. Every squad member is briefed on specific intelligence requirements (IRs) for the patrol and the unit's priority intelligence requirements (PIRs) and commander's critical information requirements as part of the patrol order. While most information will be collected during the patrol debrief, the squad leader uses judgment on what information should be reported immediately to HHQ, such as answers to PIRs.

Supervise (Continuous). Supervising is a continuous process throughout planning. Squad leaders must delegate their authority to their subordinate leaders in order to conduct coordination and planning. Inspections and rehearsals are the foundation of supervision. Squad leaders must conduct final inspections and rehearsals.

Inspections. Inspections are vital to proper preparation. They are conducted even when the squad leader and patrol members are experienced in patrolling. Inspections determine the Marines' state of readiness, both mental and physical. The squad leader conducts pre-combat checks and pre-combat inspections just prior to conducting rehearsals. During these, the squad leader looks for—

- Prescribed uniforms, weapons, ammunition, ordnance, and equipment.
- Camouflage.
- Identification tags.
- Unnecessary equipment and personal items.

The squad leader questions each member of the squad (i.e., patrol) to ensure they know—

- The mission, routes, and fire support plan.
- Their assignments and during what part of the patrol they perform them.
- What other members of the patrol are to do at certain times during the patrol.
- Challenges and passwords, call signs, frequencies, code words, reporting times, and other pertinent details.

Rehearsals. Rehearsals ensure the operational proficiency of the patrol. Plans are checked and any necessary changes are made. The patrol leader verifies the suitability of the equipment. It is through rehearsals that patrol members become thoroughly familiar with the actions they are to take during the patrol.

If the patrol is to operate at night, both day and night rehearsals should be conducted. They should be conducted on terrain similar to that on which the patrol will operate. All actions should be rehearsed. If time is limited, only the most critical actions should be rehearsed. Actions at the objective area are the most critical and should always be rehearsed.

The patrol leader should talk the squad through mission, describing the actions and having each Marine perform their duties. Rehearsals should also include a walk-through of all actions using only the signals and commands which will be used during the actual patrol.

When rehearsals are completed and the patrol leader is satisfied with the members' performance, any necessary final adjustments to the plan or patrol organization are made. Final instructions are then issued to the unit/team leaders, noting any changes made. While the subordinate leaders are giving the final instructions to their Marines, the patrol leader informs the platoon leader that the patrol is ready to depart.

> ***Note:*** If there is any extra time between the final rehearsal and the time to depart, the patrol should conduct another inspection.

TYPES OF PATROLS

Patrol missions can range from security patrols in the close vicinity of the main body to reconnaissance deep into enemy territory. The planned action determines the type of patrol. The two categories of patrols are—

- Combat.
- Reconnaissance.

Regardless of the type of patrol being sent out, a clear task and purpose must be issued to the patrol leader. Any time a patrol leaves the main body of the unit there is a possibility that it may become engaged in close combat.

Combat Patrol

A combat patrol is a tactical unit sent out from the main body to engage in independent fighting. It may be to provide security or to harass, destroy, or capture enemy troops, equipment, or installations. Combat patrols, which are typically larger and more heavily armed than reconnaissance patrols, usually depart with the intent of making contact with the enemy, followed by a return to friendly positions. Regardless of the mission, the patrol reports any information concerning the enemy and terrain that it acquires while executing the assigned mission by conducting an after action (i.e., patrol) debrief with an intelligence representative from the CLIC or the battalion S-2. There are three types of combat patrols that the squad may be tasked with: ambush, contact, and security.

Ambush. An ambush is a surprise attack by fire from concealed positions on a moving or temporarily halted enemy.

Contact Patrol. Contact patrols are combat patrols that establish and/or maintain contact to the front, flanks, or rear by contacting friendly forces at designated points; establish contact with a friendly or enemy force when the definite location of the force is unknown; and maintain contact with friendly or enemy forces.

Security. Security patrols are conducted during offense, defense, and stability activities across the full range of military operations. Just as the nature of security requirements can be different in each of the operational areas, so can the manner in which security patrols are employed.

Combat Patrol Organization. There are three essential elements (i.e., subordinate units) in combat patrols: security; support; and assault (see figure 5-1). Assault elements accomplish their missions during actions on the objective. Support elements suppress or destroy enemy on the objective in support of the assault element. Security elements assist in isolating the objective by denying possible enemy reinforcement from entering the objective area, preventing the enemy from leaving the objective area, and ensuring the patrol's withdrawal route remains open. The size of each element is based on the situation and the squad leader's analysis of METT-T. Squad leaders should seek to maintain fire team cohesion when establishing elements (e.g., assault and support). The squad leader ensures that every individual, team, and element is organized and assigned a specific task, but that ALL are capable of and prepared to perform other tasks. This may not be possible for certain specialist tasks requiring a trained technician (e.g., demolition, scout/sniper, explosive ordnance disposal).

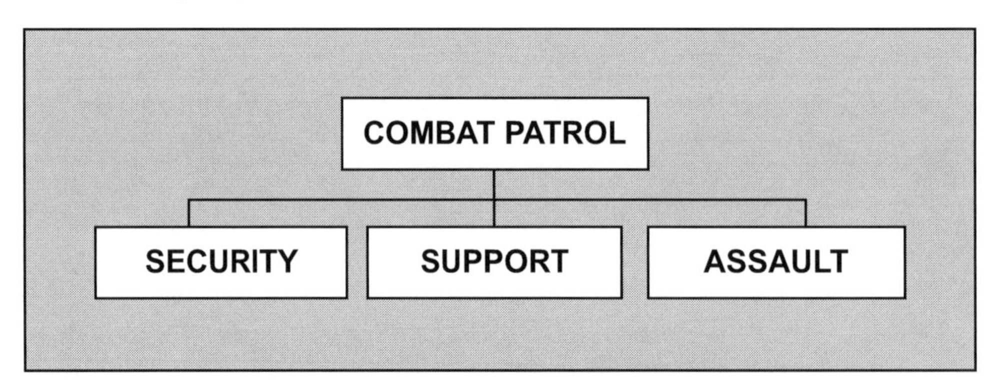

Figure 5-1. Combat Patrol Organization.

Assault. The assault element executes the squad's decisive effort (i.e., task and purpose). Its task is to conduct the necessary actions on the objective. This element must be able to destroy or seize the target. Tasks typically associated with the assault element include—

- Conducting an assault across the objective to destroy enemy equipment, capture or kill enemy personnel, or clear key terrain and enemy positions.
- Deploying close enough to the objective to conduct an immediate assault if detected.
- Being prepared to support itself if the support element cannot suppress the enemy.
- Providing support to an attached breach element to reduce obstacles, if required.
- Conducting a controlled withdrawal from the objective.

Marines within the assault element can expect to be augmented with or tasked to fill the following special teams:

- *Search teams*. Search teams are tasked with searching the objective area for documents, equipment, or information that may be of intelligence value.
- *Detainee teams*. Detainee teams are tasked to capture, secure, or detain enemy forces or possible adversaries.
- *Demolition teams*. Demolition teams plan and execute the destruction of obstacles and enemy equipment.
- *Aid and litter teams*. Aid and litter teams are tasked to identify and collect casualties, administer immediate aid, and coordinate CASEVAC.

Support. The purpose of the support element is to suppress the enemy on the objective using direct and indirect fires. The support element is a shaping effort that sets conditions for the mission's decisive effort (i.e., the assault element). The support force may be divided into two or more elements if required, based on METT-T.

The support element must be organized to contend with any secondary enemy threat that could interfere with the assault elements. The support elements suppress, fix, or destroy enemy elements on the objective. The support element's primary duty is to suppress enemy forces; preventing the enemy from repositioning against the assault elements. The support force—

- Gains fire superiority with crew-served weapons and indirect fires.
- Controls the rate and distribution of fires.
- Shifts or ceases fire on signal.
- Supports the withdrawal of the assault element.

Security. The security element performs four roles:

- It isolates the objective area from enemy forces attempting to enter it.
- It prevents enemy forces from escaping the objective area.
- It protects the assault and support elements.
- It secures the patrol's withdrawal route.

The security element fills a critical task. While all elements of the patrol are responsible for their own local security, what distinguishes the security element is that it protects the entire patrol. Their positions must be such that they can—depending upon the engagement criteria—provide early warning of approaching enemy forces. To facilitate the success of the assault element, the security element must fix or block all enemy response forces away from the objective area.

Ambush Patrol

An ambush patrol does not need to seize or hold terrain. It may close with and destroy the enemy (i.e., near ambush) or attack by fire (i.e., far ambush). Ambushes must be executed with ferocity and violence of action, seeking decisive effects in the opening moments of the engagement. The size of the enemy, the orientation of the ambush site, the battlespace geometry, security, and the size of the ambush element generally dictate what means of employment the squad leader chooses. Leaders execute ambushes to reduce the enemy's overall combat effectiveness for the specific purpose of destroying its units. Destruction is the primary purpose of an ambush because the loss of personnel, equipment, or supplies reduces the enemy's overall combat effectiveness. Frequent ambushes force the enemy to divert soldiers from other missions to guard convoys, troop movements, and logistics resupply efforts.

Based on the amount of time available, there are two ambush options:

- *Hasty ambush*. A hasty ambush is based on an unforeseen opportunity. It is used when a patrol sees the enemy before the enemy sees them, and the patrol has time to act, usually as an immediate action drill.
- *Deliberate ambush*. A deliberate ambush is conducted against a specific target at a location chosen based on intelligence. With a deliberate ambush, leaders plan and prepare based on detailed information that allows them to anticipate enemy actions and enemy locations. For more on hasty and deliberate ambushes, see MCTP 3-01A.

Ambush Terminology. During METT-T analysis, the squad leader identifies four different locations related to ambushes: the kill zone, the ambush site, security positions, and rally points. Squad leaders should consider the following when selecting these four positions.

The ambush site is the terrain on which a point ambush is established. An ambush site contains a support by fire position for the support element and an assault position for the assault element. The ambush site should not be in an obvious location. It should provide—

- Good fields of fire into the kill zone.
- Good cover and concealment.
- Protective obstacles (e.g., a river, a canal) between the ambush site and kill zone, if possible.
- A covered and concealed withdrawal route.
- A defense against a possible enemy flank attack.

The kill zone is the part of the ambush site where concentrated fire is delivered to trap, isolate, or destroy enemy forces. The concentration of fire may vary depending upon whether it is a near or far ambush. The kill zone should have the following characteristics:

- A location that enemy forces are likely to enter.
- Provide natural tactical obstacles between the kill zone and the assault position, when possible.
- Large enough to observe and engage enemy forces.

A near ambush is a point ambush where the assault element is within reasonable assaulting distance of the kill zone (i.e., less than 50 meters). Close terrain, such as an urban area, jungle, or heavy woods, may require this positioning.

A far ambush is a point ambush where the assault element is beyond reasonable assaulting distance of the kill zone (i.e., beyond 50 meters). This location may be appropriate in open terrain offering good fields of fire, or when attack is by fire for a harassing ambush.

When selecting security positions, the squad leader should consider the following:

- Positions should not mask fires from the support element or assault force.
- Provide timely information to the squad leader to allow enough time to act on the information.
- They should provide a possible support by fire position, if required.

The squad leader considers the use and locations of rally points. Rally points are set where the squad will move to reassemble and reorganize if it becomes dispersed. Selected during route reconnaissance whenever possible, rally points should—

- Be easy to recognize on the ground.
- Provide cover and concealment.
- Be away from natural lines of drift.
- Be defensible for short periods of time.

Types of Ambushes. The two types of ambushes are point and area. In a point ambush, the squad is deployed to attack a single kill zone (see figure 5-2). In an area ambush, the squad would typically be part of a larger force, usually the platoon or company, deployed in two or more associated point ambushes (for more on area ambushes, see MCTP 3-01A). These ambushes at separate sites are related by their purposes.

*Ambush Formations.*There are many possible ambush formations. Due to the squad's limited size, those most common in a point ambush are the L-shaped and V-shaped ambush formations.

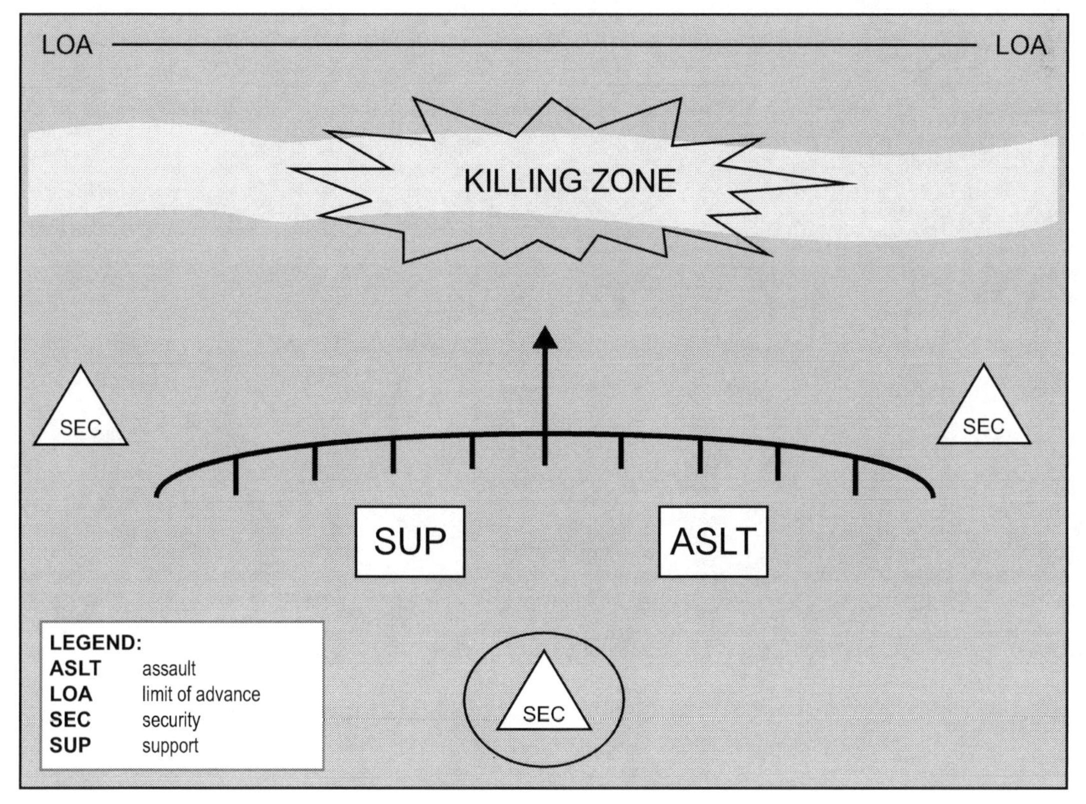

Figure 5-2. Point Ambush.

For more information on ambush formations, see MCTP 3-01A. All formations require leaders to exercise strict direct fire control. Leaders need to understand the strengths and weaknesses of their units and plan accordingly. The following should be considered when selecting formations:

- Terrain.
- Visibility.
- Marines available.
- Weapons and equipment.
- Ease of control.
- Target to be attacked.

An ambush in the L-shaped formation is a variation of the linear formation (see figure 5-3). The long leg of the L (i.e., assault element) is parallel to the kill zone and provides flanking fire. The short leg (i.e., support element) is at the end of and at a right angle to the kill zone, and provides enfilading fire that works with fire from the other leg. The L-shaped formation is best employed at a sharp bend in a trail, road, or stream.

In a V-shaped ambush, assault elements are placed along both sides of the enemy's route so they form a V (see figure 5-4). Extreme care must be taken to ensure that neither element fires upon the other. This formation subjects the enemy to both enfilading and interlocking fire.

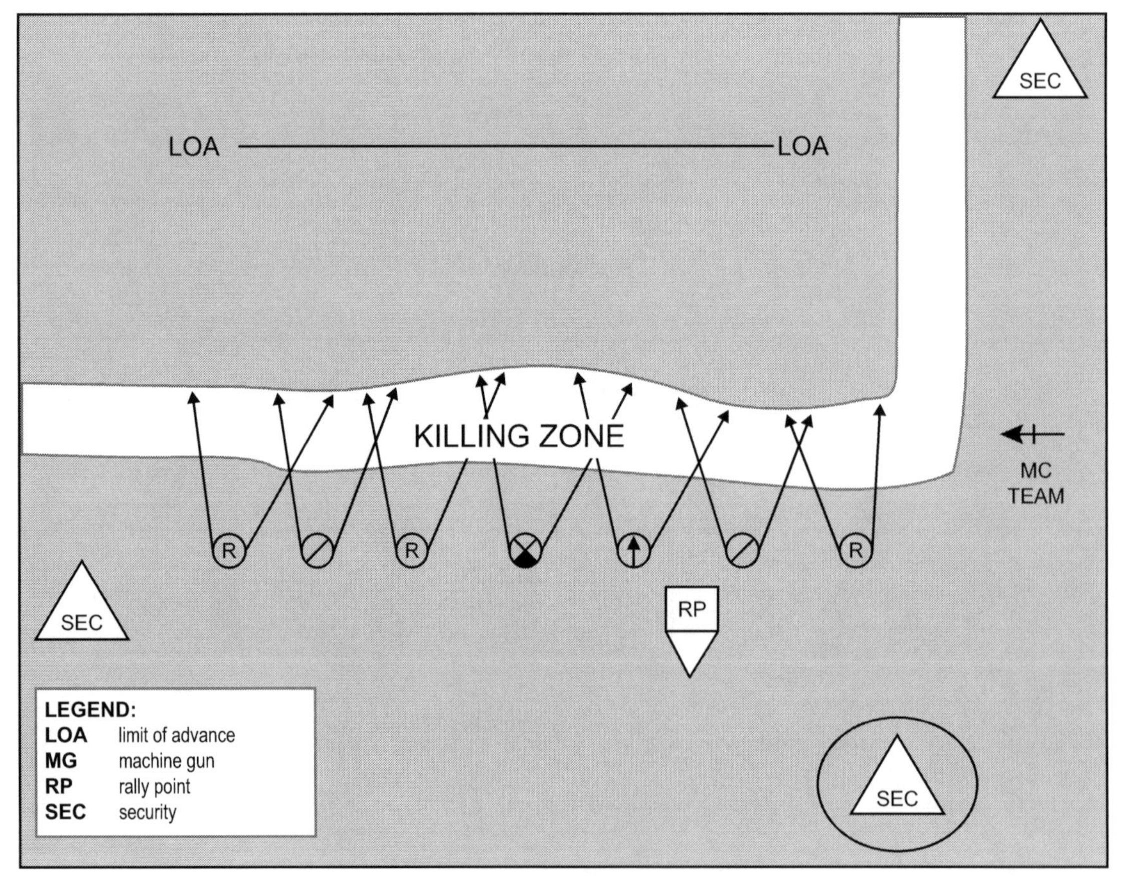

Figure 5-3. L-Shaped Ambush Formation.

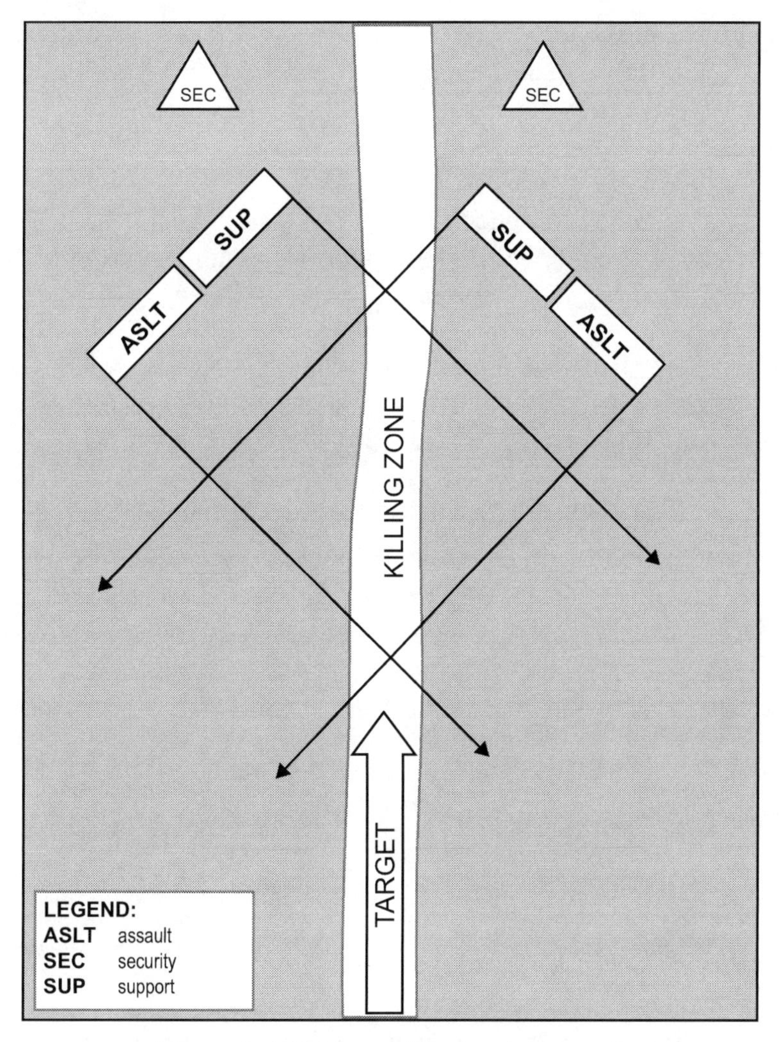

Figure 5-4. V-Shaped Ambush Formation.

Wider separation of the elements makes this formation difficult to control, and there are fewer sites which favor its use. Its main advantage is that it is difficult for the enemy to detect the ambush until they are well into the kill zone. As stated earlier, in a point ambush, the squad is deployed to attack an enemy target in a single kill zone. The squad leader usually stays with the assault element for control. The squad leader conducts a leader's reconnaissance of the objective prior to sending elements forward. Prior to departing, a five-point contingency plan is issued in case of unforeseen situations as follows:

G	Going: Where is the leader going?
O	Others: what others are going with the leader?
T	Time (duration): how long will the leader be gone?
W	What: what do we do if the leader fails to return?
A	Actions: what actions do the departing reconnaissance element and main body plan to take on contact?

Figure 5-5. Leader's Five-Point Contingency Plan.

The squad leader confirms the condition of the objective, gives each subordinate leader a clear picture of the terrain where they will move, and identifies all locations within the ambush site. The leader's reconnaissance can consist of the squad leader, element leaders, and (sometimes) security personnel. It gets back to the patrol as quickly as possible.

The security elements should be positioned first. The support element should then be emplaced before the assault element moves forward. The support element must provide overwatch for the movement of the assault element into position.

The squad leader must check each member of the assault element once emplaced. The squad leader signals the support and security elements once the assault element is emplaced. The actions of the assault element, support element, and security element are depicted in table 5-1.

Table 5-1. Actions of Ambush Elements.

Assault Element	Support Element	Security Element
Identify individual sectors of fire the squad leader assigned; emplace aiming stakes. Emplace protective obstacles. Emplace command detonated explosives in dead space within the killing zone. Camouflage positions. Take weapons off safe when directed by the squad leader.	Identify sectors of fire for all weapons, especially machine guns. Emplace limiting stakes to prevent friendly fires from hitting the assault element as required. (e.g. L or V -shaped ambush). Emplace command detonated explosives and other protective obstacles. Camouflage positions.	Identify sectors of fire for all weapons; emplace aiming stakes. Emplace command detonated explosives and other protective obstacles. Camouflage positions. Secure the ORP; secure a route to the ORP as required. Cover the withdrawal of the assault and support elements.

The security element (or team) notifies the squad leader of the enemy's approach into the killing zone using a pre-arranged signal. The security element must also alert the squad leader if any additional enemy forces are following the lead enemy force. For this reason, security elements must be positioned so that constant communications can occur with the squad leader. This allows the squad leader to know if the enemy force meets the engagement criteria. The squad leader must be prepared to allow free passage to enemy forces that are too large or that do not meet the engagement criteria. Any enemy forces that pass through the ambush site without being engaged must be reported to HHQ utilizing the proper report format in the unit's SOP.

The squad leader initiates the ambush by opening fire and shouting, *Fire!* and thus ensuring initiation of the ambush should the rifle fail to fire. The squad leader must also have signals to cease fire, shift fire, assault or conduct actions in the killing zone, and withdraw. All Marines in the ambush must understand all signals. If the patrol is detected before this, the first member to become aware initiates the ambush by firing and shouting. The platoon should rehearse with both methods to avoid confusion or the loss of surprise when executing the ambush.

> ***Note:*** The squad leader must plan to engage the enemy in limited visibility. Based on the unit's SOP, the squad leader should consider using a mix of tracers and the employment of illumination, pyrotechnics, and NVDs.

The squad leader should also include the employment of indirect fire support in the plan. Based on the platoon commander's guidance, the squad leader may employ indirect fires to cover the flanks of the killing zone to isolate an enemy force or to assist with disengagement if the ambush is compromised or if the squad must depart the ambush site under pressure.

The squad leader must have a solid plan (i.e., both day and night) to signal the security and support elements when the assault element is moving into the kill zone to begin its search and site exploitation activities. Existing environmental factors must be considered. For example, smoke for obscuration may not be visible to the support element because of limited visibility or the lay of the terrain. All Marines must know and practice relaying signals during rehearsals to avoid the potential for fratricide.

The support element must be prepared to move across the killing zone and establish far side security for the search teams. Once the search of the killing zone is complete or on order from the squad leader, the search and far side security teams withdraw in reverse order. Search techniques in the killing zone should be in accordance with unit SOP.

If a security team makes enemy contact during the squad's withdrawal, it fights as long as possible without becoming decisively engaged using pre-arranged signals and informs the squad leader it is breaking contact. The squad leader may direct a portion of the support element to assist the security element in breaking contact.

The withdrawal plan from the ambush site should include the following:

- Elements are typically withdrawn in the reverse order from occupation.
- Elements may return to a release point, then to the ORP, depending on the distance between the elements.
- The security element at the ORP must be alert to assist the platoon's return. It maintains security for the ORP while the remainder of the platoon prepares to depart.
- Accounting for personnel and equipment, packing captured enemy equipment, applying first aid, and transporting casualties (as necessary).

Contact Patrol

A contact patrol is a patrol sent out from one unit to make contact with another unit, either friendly or threat forces. Though modern technology has reduced (but not eliminated) the need for contact patrols to be conducted between US forces, there still may be a need when a US force must contact a non-US coalition partner which lacks compatible communications or position reporting equipment. Contact patrols may go to the other unit's position, or the units can meet at a designated contact point. The squad leader provides the unit with information about the location, situation, and intentions of their own unit, and obtains and reports the same information about the contacted unit back to HHQ.

The contact patrol also observes and reports pertinent information about the area between the two units. Though this is one possible mission of a contact patrol, the standard mission of a contact patrol is to maintain contact to the front, flanks, or rear of the unit by—

- Establishing contact with a threat force when its precise location is unknown.
- Maintaining contact with enemy forces through the direct or indirect effects of fires, or through observation.
- Avoiding becoming decisively engaged with the enemy.

Security Patrol

A security patrol can be conducted during offense, defense, and stability activities across the full range of military operations. Just as the nature of security requirements can be different in each of the operational area, so can be the manner in which security patrols are employed. Security patrols meet usual internal requirements of the parent unit by—

- Protecting its front, flanks, rears, areas, and routes.
- Protecting static positions from infiltration.
- Providing early warning of enemy attack.
- Disrupting threat reconnaissance efforts and preventing surprise.

Security patrols do not operate beyond the range of communication and supporting fires from the main body, especially from mortar fires, because they typically operate for limited periods of time and are combat oriented.

In rear areas, particularly when conducting stability activities in a nonpermissive environment, the requirement to conduct security patrols increases for all MAGTF units ashore, particularly for aviation and combat service support units.

> ***Note:*** All procedures pertaining to the patrol activities presented above are to be utilized in security patrols.

Reconnaissance Patrol

Reconnaissance patrols gather information about the enemy, terrain, or resources for the purpose of confirming or disproving the accuracy of information. The squad is ideally suited for reconnaissance patrol missions because of its relatively small size, cohesion, and experience working together. Reconnaissance patrols are conducted to gather information on—

- The locations of possible threat forces, installations, and equipment.
- The identification of enemy units and equipment.
- The strength and disposition of enemy forces.
- The movement of enemy personnel and equipment.
- New or special types of weapons.
- Unusual activities of threat forces.
- Patterns of life.
- Population atmospherics.

The goal of reconnaissance patrols is to rely on stealth rather than combat strength, to gather the necessary information, and to fight only when necessary to complete the mission or to defend themselves. A reconnaissance patrol typically travels light, with only the required personnel, arms, ammunition, and equipment necessary to complete the mission. This increases stealth and cross-country mobility in close terrain. Regardless of how the patrol is armed and equipped, reconnaissance patrols must be able to rapidly transition to combat. For more information, see MCRP 2-10A.6, *Ground Reconnaissance Operations*.

The three types of reconnaissance patrols include:

- Area. An area reconnaissance is a directed effort to obtain detailed information concerning the terrain or enemy activity within a prescribed area such as a town, ridge line, woods, or other features critical to operations.
- Route. A route reconnaissance is a directed effort to obtain detailed information on a specified route and all terrain from which the enemy could influence movement along that route.
- Zone. A zone reconnaissance is a directed effort to obtain detailed information concerning all routes, obstacles (to include chemical or radiological contamination), terrain, and enemy forces within a zone defined by boundaries. A zone reconnaissance typically is assigned when the enemy situation is vague or when information concerning cross-country trafficability is desired.

Information Requirements. Information requirements provide commanders information needed to make decisions. Some IRs are the commander's critical information requirements. It is the responsibility of the squad's HHQ to clearly define the IRs they want the patrol to answer, and the responsibility of the squad leader to clarify the IRs prior to conducting the mission.

Observation Plan. Once squad leaders understand the required IRs, they determine how the patrol can best obtain them by developing an observation plan. Squad leaders develop their plans by asking two questions:

- What are the best locations to obtain the information required?
- How can the information best be obtained without compromising the patrol?

The answer to the first question is by selecting OPs based on METT-T analysis that will allow clear fields of view on the objectives and occupying them only until the IRs have been answered, confirmed, or denied. The answer to the second question is to use only the number of routes and number of personnel needed to occupy the OPs to observe the objectives.

In establishing the number and locations of OPs, the squad leader should consider whether the positions provide the following characteristics:

- Covered and concealed routes to and from each OP.
- Unobstructed observation of the objectives.
- Effective cover and concealment to reduce vulnerability and increase survivability.

- A location that does not attract attention, away from natural lines of drift or prominent terrain features (e.g., a stand of trees or a water tower).
- A location that does not skyline the observers.

Reconnaissance Patrol Organization. While the organization of reconnaissance patrols is METT-T dependent, they are usually organized around the current structure of the Marine rifle squad. One or more of the squad's fire teams is assigned as the reconnaissance element to reconnoiter or maintain surveillance over the objective, and at least one fire team is assigned as a security element. In a squad-size patrol, the headquarters may make up part of one of the two elements. Reinforcing the squad is based on METT-T analysis. Figure 5-6, figure 5-7, and figure 5-8 show examples of how a squad leader can organize a reconnaissance patrol.

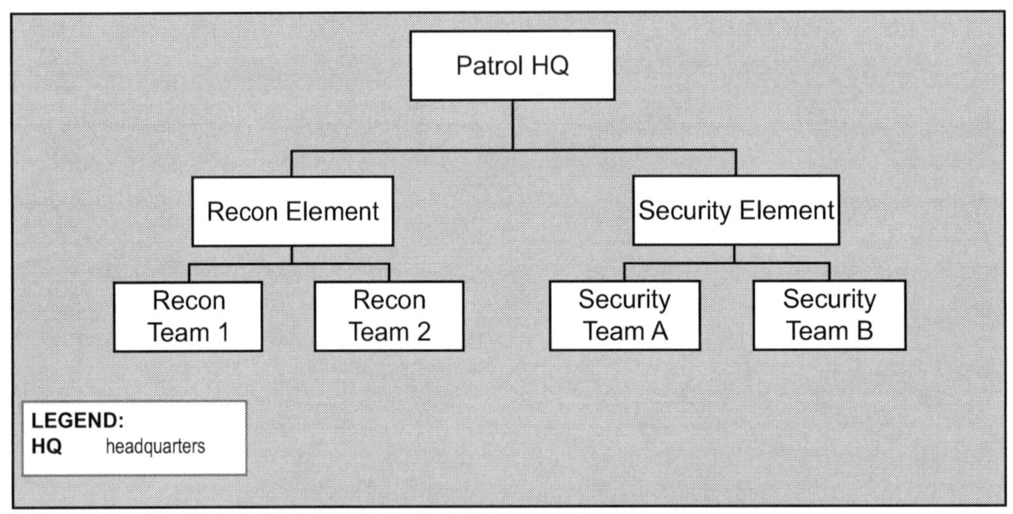

Figure 5-6. Reconnaissance Patrol Organization.

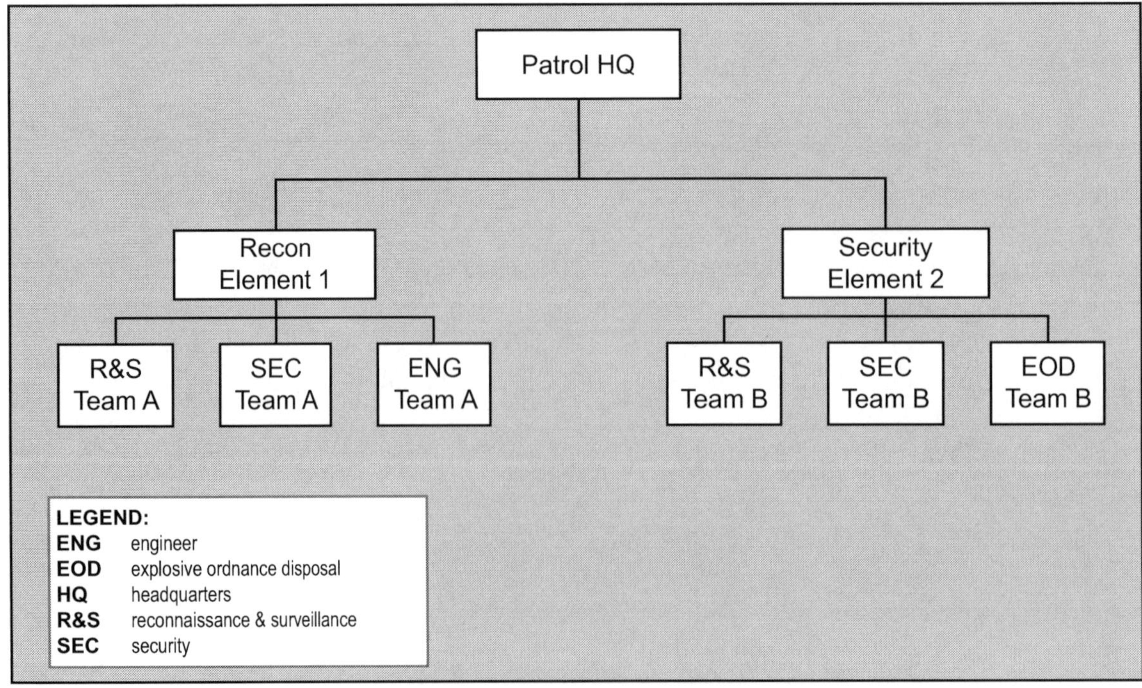

Figure 5-7. Reconnaissance Patrol Organization (with enablers).

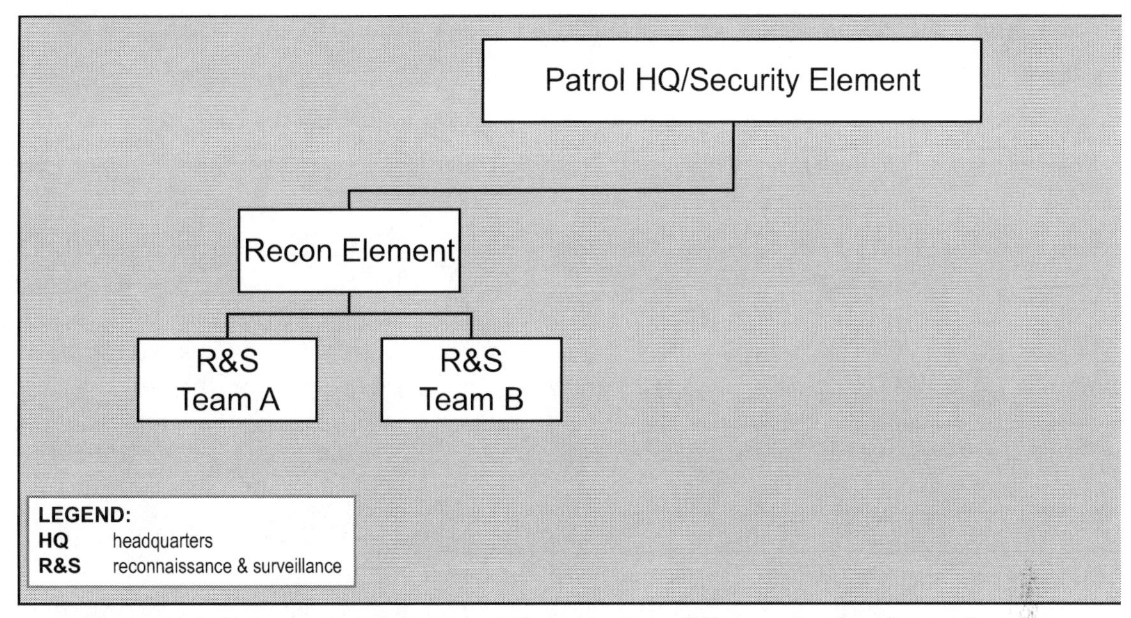

Figure 5-8. Reconnaissance Patrol Organization (HQ as security element).

Reconnaissance Element. The purpose of the reconnaissance element is to answer IRs for the purpose of facilitating tactical decision making. This is primarily accomplished through reconnaissance (or surveillance) and continuous, accurate reporting. The squad leader determines how in-depth the reconnaissance will be, based on the clear purpose and task given by HHQ. A thorough and accurate reconnaissance is important, but so is avoiding detection.

The following are additional tasks associated with the reconnaissance element, either by a rifle squad or with augmentation from subject matter experts such as engineers, civil affairs, counterintelligence, or communicators. Some of these tasks include—

- Determine the trafficable routes or potential avenues of approach based on the personnel or vehicles to be used on the route. Subsets of this may include—
  - Inspect and classify all bridges, overpasses, underpasses, and culverts on the route.
  - Locate fords or crossing sites near bridges on the route.
- Reconnoiter to the limit of direct fire range:
  - Terrain that influences the area, route, or zone.
  - Built-up areas.
- Reconnoiter natural and man-made obstacles to ensure mobility along the route.
- Locate a bypass or reduce/breach, clear, and mark:
  - Lanes.
  - Defiles and other restrictive terrain.
  - Minefields.
  - Contaminated areas.
  - Log obstacles such as abatis, log cribs, stumps, and posts.
  - Wire entanglements.
  - Other obstacles along the route.

- Determine the size, location, and composition of social/human demographics.
- Identify key infrastructure that could influence military operations, including the following:
  - Political, government, and religious organizations and agencies.
  - Physical facilities and utilities (e.g., power, transportation, communications networks).

Security Element. The security element has two main tasks: provide early warning, and provide support by fire to the reconnaissance element. The purpose of the security element is to enable the reconnaissance element to obtain answers to HHQ's IRs. The security element must be positioned to observe avenues of approach into and out of the objective area. If the reconnaissance element is compromised, the security element must be able to quickly support it. It does so by occupying positions allowing it to observe the objective as well as cover the reconnaissance element. It must also be able to facilitate communication to HHQ, as well as any supporting assets.

Area Reconnaissance. In an area reconnaissance, the squad may use several surveillance points, vantage points, or OPs around the objective to observe it and the surrounding area (see zone reconnaissance). Actions at the objective for an area reconnaissance usually begin with the squad in the ORP, and end with the dissemination of collected information after a linkup of the reconnaissance/surveillance elements in the ORP. The critical actions include—

- Actions from the ORP.
- Execution of the observation plan.
- Linkup of reconnaissance elements.

The squad occupies the ORP, establishes local security, and conducts associated mission preparations. While the squad prepares for the mission, the squad leader and selected personnel conduct a leader's reconnaissance. Three things must be accomplished during this reconnaissance:

- Pinpoint the objective.
- Identify a release point and follow-on linkup points (if required).
- Confirm the observation plan.

Upon returning from the leader's reconnaissance, the squad leader issues information and tasks as required. Once ready, the squad departs. The reconnaissance elements move along the designated routes to the OPs in accordance with the observation plan.

Once in position, the reconnaissance elements observe and listen to acquire the needed information. No eating, talking, or unnecessary movement occurs at this time. If the reconnaissance element cannot acquire the information needed from its initial position, it retraces the route and repeats the process. This method of reconnaissance is extremely risky.

> ***Note:*** The reconnaissance element must remember that the closer it moves to an objective, the greater the risk of being detected.

Route Reconnaissance. Route reconnaissance is oriented on a road, a narrow axis such as an infiltration lane, or on a general direction of attack. Squads conducting route reconnaissance missions should view the route from both the friendly and enemy perspectives. Infantry squads require augmentation with technical experts (e.g., combat engineers) for a complete detailed route reconnaissance. However, squads should be capable of conducting a quick route reconnaissance or area reconnaissance of selected areas along a route. The squad could also be tasked to survey a route along a planned infiltration lane. After being briefed on the proposed infiltration, the squad leader conducts a thorough map reconnaissance and develops the surveillance plan. The platoon reports conditions likely to affect friendly movement. These conditions include—

- The presence of the enemy.
- Terrain information.
- The location and condition of bypasses, fords, and obstacles (both man-made and natural).
- Chokepoints.
- Route and bridge conditions.

> ***Note:*** If all or part of the proposed route is a road, the squad should treat the road as a danger area. The squad conducts the reconnaissance by moving parallel to the road, using a covered and concealed route. When required, reconnaissance and security teams move close to the road to reconnoiter key areas. The platoon plans a different route for its return. Upon the squad's return, the squad leader may be tasked to submit a report in an overlay format. For more information on reconnaissance reports, see MCRP 2-10A.7, *Reconnaissance Reports Guide*.

Zone Reconnaissance. A zone reconnaissance is conducted to obtain information on the enemy, terrain, and routes within a specified zone. Zone reconnaissance techniques include the use of moving elements, stationary teams, or a combination of both. When moving elements are required, squads and/or fire teams move along multiple routes to cover the whole zone. When the mission requires the squad to flood an area, the squad may use one of the following methods: the fan, converging routes, or successive-sector.

Fan. When using the fan method, the squad leader selects a series of ORPs throughout the zone to operate from. The patrol establishes security at the first ORP; upon confirming the ORP location, the squad leader confirms routes out from and back to it. These routes form a fan-shaped pattern around which the ORP is the hub. The routes must overlap to ensure the entire area is reconnoitered. Once the routes are confirmed, the leader sends out reconnaissance teams along the routes. Each reconnaissance team moves from the ORP along a different route, or blade of the fan, that overlaps with others to ensure reconnaissance of the entire area (see figure 5-9). When all teams have returned to the ORP, the squad leader collects and disseminates all information gathered to every Marine and to HHQ before moving on to the next ORP.

Converging Routes. When using the converging routes method, the squad leader sends the reconnaissance teams from the ORP along routes that surround the objective. The teams may occupy multiple OPs around the objective in order to answer the required RFIs. Once complete,

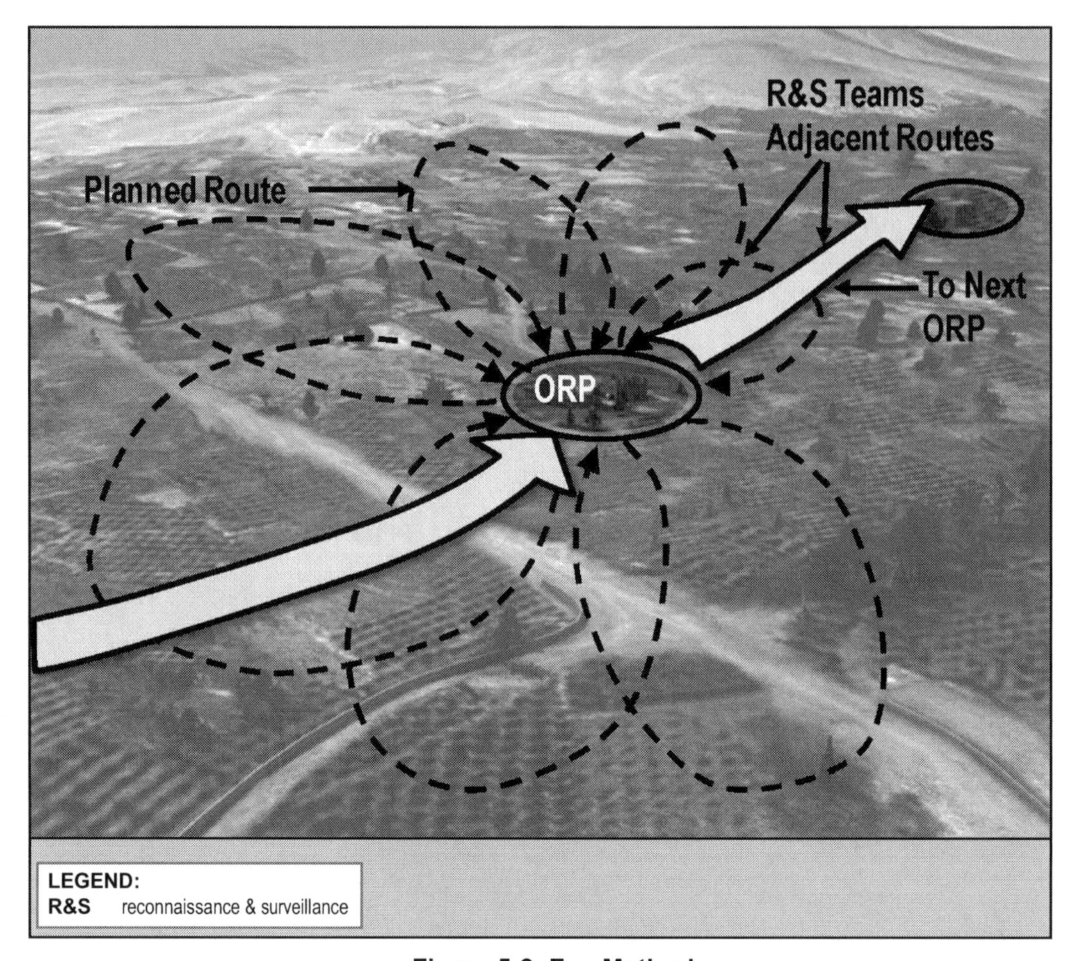

Figure 5-9. Fan Method.

the teams move to the pre-designated rally point, which is usually located on the far side of the objective (see figure 5-10).

Successive-Sector. This method is a continuation of the converging routes method. The squad leader selects an ORP, a series of reconnaissance routes, OPs, and rally points. The actions of each team from ORP to rally point are the same as for the converging routes method; that is, each rally point becomes the ORP for the next phase. After linkup, the squad leader issues new routes, OPs, linkup time, and the next rally point. These actions continue until the entire zone has been reconnoitered (see figure 5-11).

IMMEDIATE ACTIONS UPON ENEMY CONTACT

The squad can make contact with the enemy at any time through observation, a meeting engagement, or an ambush. Contact may be visual, where the squad sights the enemy but remains undetected. In this case, the squad leader can decide whether to initiate or avoid physical contact, based on the squad's mission and capability to effectively engage the enemy force. The following actions may be performed, depending on the type of patrol and enemy contact:

- Reconnaissance patrol. The reconnaissance patrol's mission should prohibit physical contact unless necessary to accomplish the mission. The squad's actions upon contact are usually defensive in nature. If contact is inevitable, it must be broken as quickly as possible, and the squad should continue the mission if it is able.

- *Combat patrol*. Since the combat patrol's purpose is to seek or exploit opportunities for contact, actions upon contact are usually offensive in nature. When enemy contact is made, the squad's actions must be swift and violent in an effort to inflict maximum damage on the enemy force, followed by immediately relocating to another area or returning to friendly lines. Patrols can expect to make physical contact with the enemy in one of two ways:
  - Meeting engagement: A meeting engagement is a combat action that occurs when a moving force, incompletely deployed for battle, engages an enemy at an unexpected time and place. It is an accidental meeting where neither the patrol nor the enemy expect contact and are not specifically prepared to deal with it.
  - Ambush: (See the various ambush discussions within this chapter.)

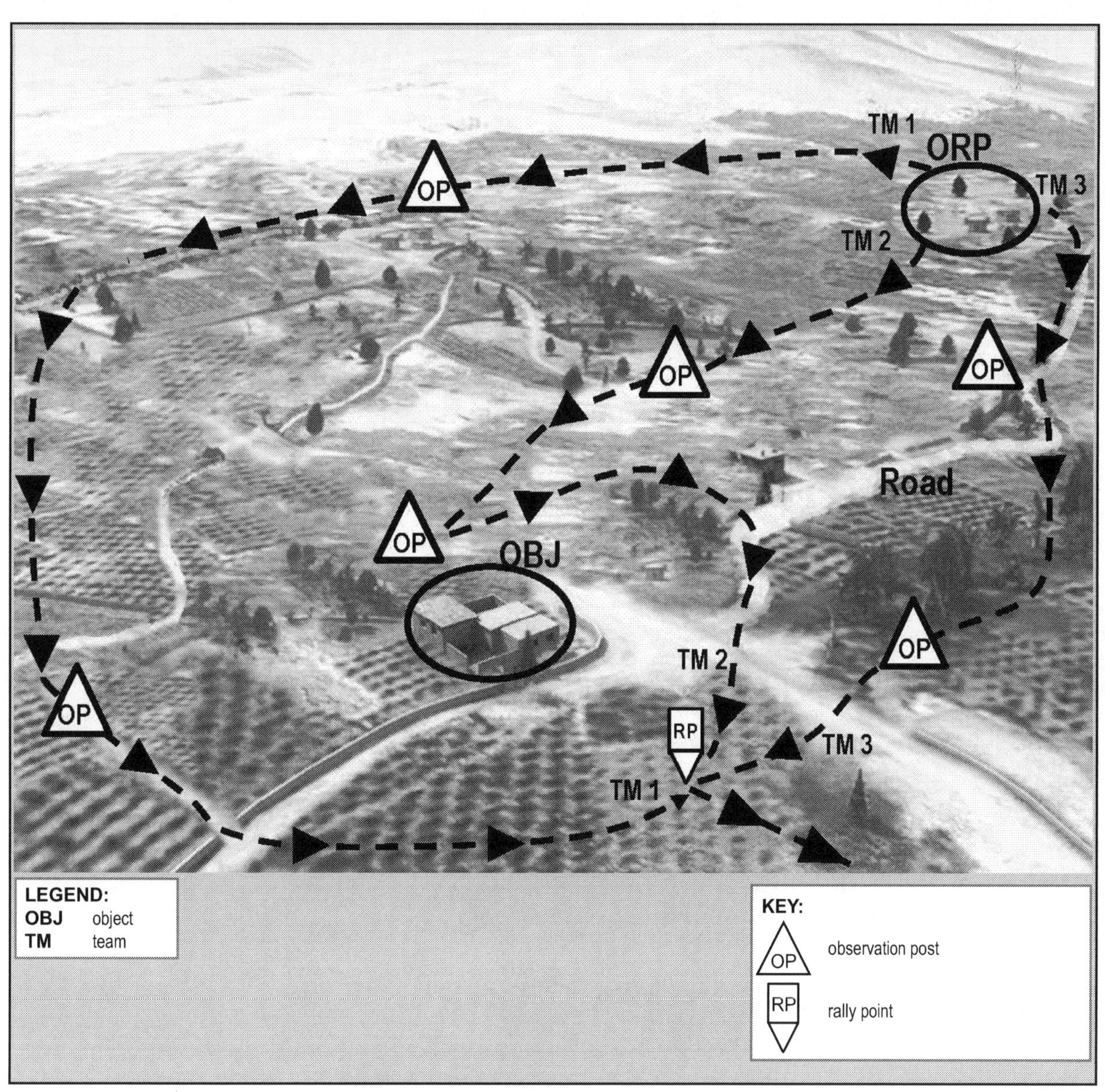

Figure 5-10. Converging Routes Method.

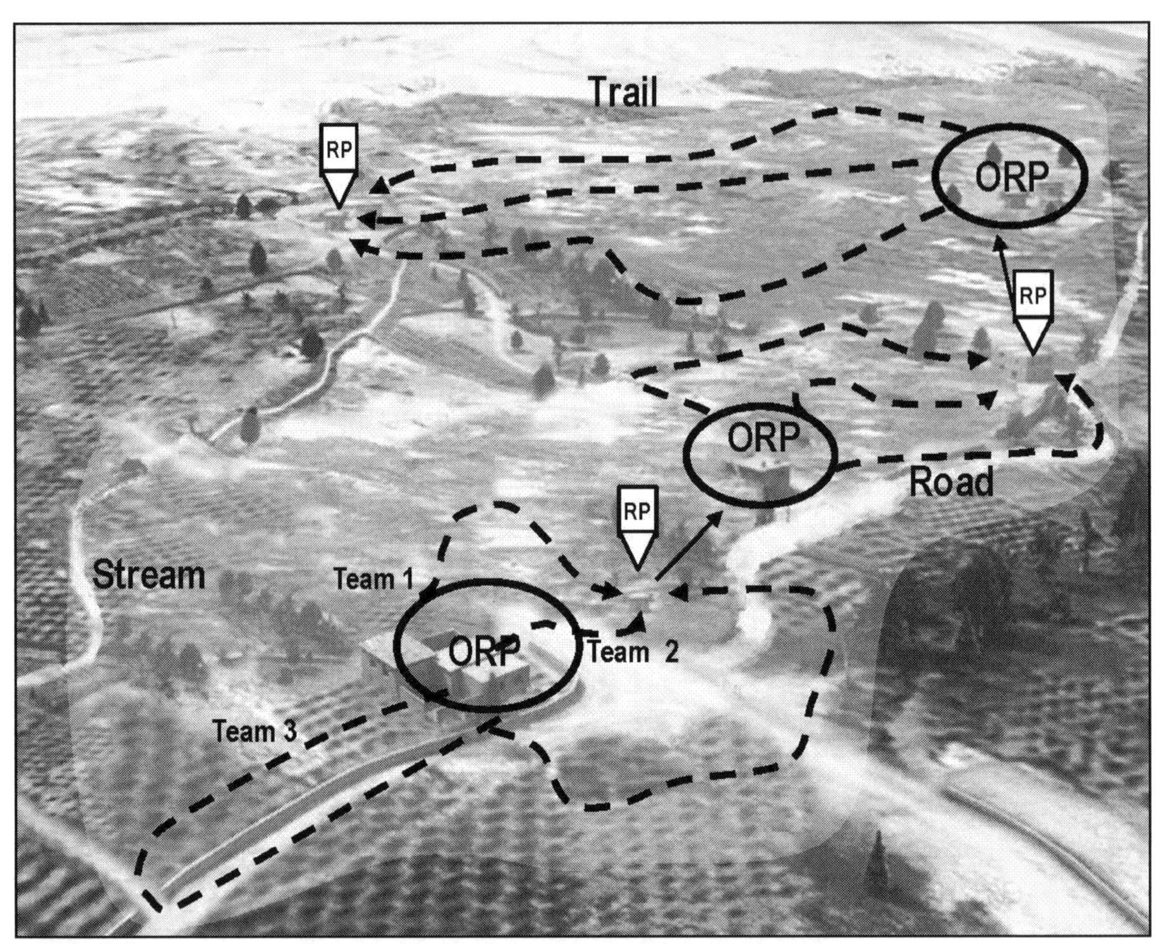

Figure 5-11. Successive-Sector Method.

Immediate Action Drills

During patrols, contact—visual or physical—is often unexpected, at very close range, and of short duration. Effective enemy fire often allows leaders little or no time to evaluate the situation and issue orders. In these situations, Marines rely on prior training to swiftly initiate positive offensive or defensive action, as appropriate. Immediate action drills are simple courses of action standardized throughout the unit, that are repeatedly rehearsed until they become second nature. It is not feasible to attempt to develop an immediate action drill to cover every possible situation, but they are extremely useful for the most common or dangerous situations. Hand-and-arm signals associated with the following common immediate action drills can be found in appendix B.

Immediate Halt. When the patrol detects the enemy but is not detected itself, the situation requires the patrol to immediately halt in place. The first member visually detecting the enemy gives the hand-and-arm signal for FREEZE. Every member halts in place and remains absolutely motionless and quiet until further signals or orders are given.

Air Observation/Air Attack. These actions are designed to reduce the danger of detection by aircraft and limit casualties from air attack.

Air Observation. When the patrol is operating in an area that provides cover and/or concealment (i.e., woodland environment) and a member of the patrol hears or observes unidentified aircraft or a small UAS that may detect the patrol, the hand-and-arm signal to FREEZE is given and all Marines should do the following (see figure 5-12):

- Halt in place until the patrol leader identifies the aircraft and gives further signals or orders.
- Immediately seek any cover available within the immediate vicinity when operating in an open area with little to no cover/concealment (e.g., desert environment).
- Maximize the use of micro-terrain or available concealment to blend in with surroundings.
- Marines must not look up, as sunlight can reflect off their faces, even when camouflaged.

Air Attack. When an attack aircraft or armed UAS detects the squad and makes a low-level attack, the immediate action drill for air attacks is used. The first member sighting an attacking aircraft shouts, *Aircraft!*, followed by the direction of the incoming attack: *front, left, rear* or *right*. Then, the squad does the following:

- The Marines move quickly to cover and stop unnecessary movement.
- Between strafing runs by the aircraft, patrol members seek better cover and a means to egress or eliminate the threat.
- Attacking aircraft are fired upon only on the patrol leader's command.

Figure 5-12. Immediate Action for Aerial Observation.

Meeting Engagements. The immediate actions described in the following subparagraphs are common reactions to a meeting engagement with enemy forces.

Hasty Ambush. A hasty ambush is the immediate action executed to avoid contact and/or execute a rapid, unplanned ambush on an unsuspecting enemy force. It often follows the command, FREEZE. On the signal for the hasty ambush, the entire patrol moves quickly to the right or left of the line of movement, as indicated by the squad leader, and takes up the concealed firing positions (see figure 5-13). The squad leader conducts the same actions as for a point ambush.

Immediate Assault. This immediate action drill is used either defensively to make and quickly break undesired but unavoidable contact (including near ambush), or to offensively engage the enemy (including far ambush). In a meeting engagement, the members nearest the enemy open fire and shout, *Contact!,* followed by the direction of the incoming attack: *front, left, rear*, or *right*, followed instantly by a rehearsed reaction. When used offensively, the enemy is decisively engaged; escapees are pursued and destroyed until the patrol leader gives orders to break contact. Engaging any longer than necessary can jeopardize the mission. The defensive assault ends if the enemy withdraws and contact is broken. If the enemy stands fast, the assault is carried through the enemy positions and movement is continued until contact is broken.

> ***Note:*** Extreme discipline and caution must be exercised so as not to be baited into an organized enemy ambush.

In a near ambush (i.e., within 50 meters or hand grenade range), Marines in the killing zone immediately return fire, take up covered positions, and throw fragmentation or smoke grenades.

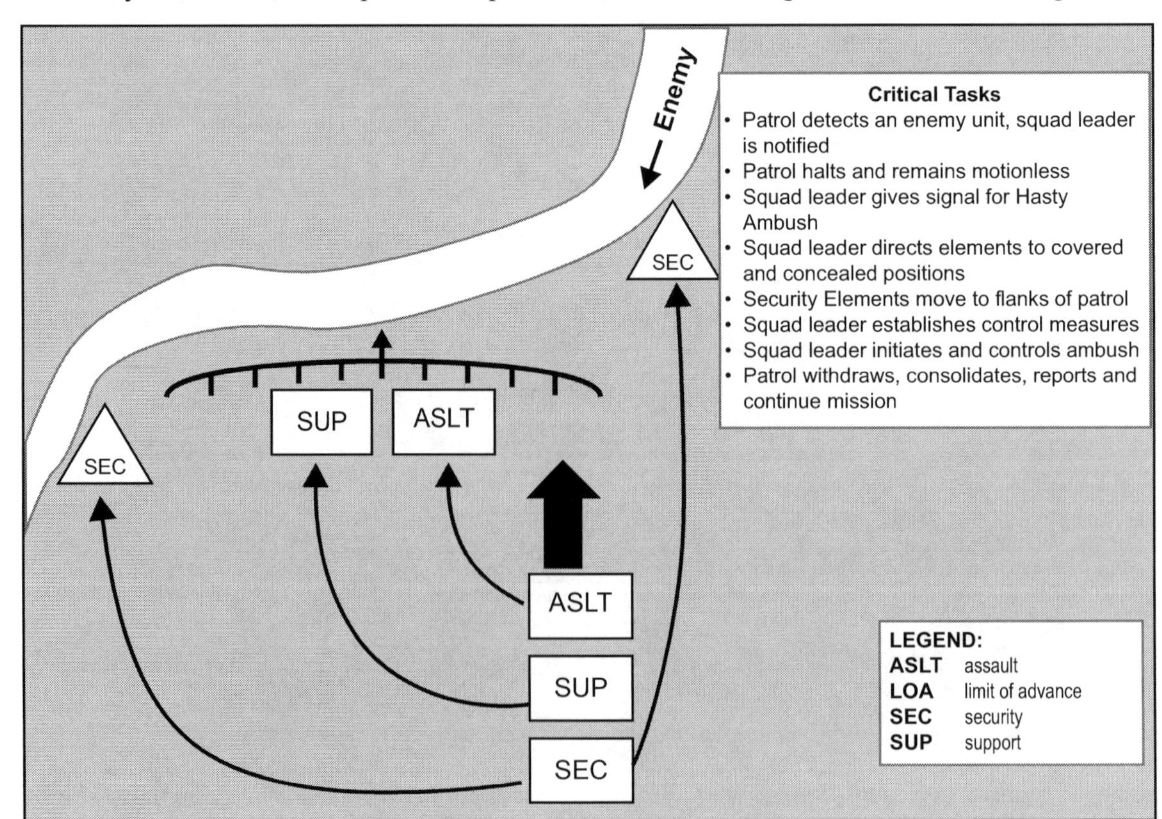

Figure 5-13. Hasty Ambush.

On grenade detonation, Marines assault through the ambush using fire and movement. Marines not in the killing zone identify enemy positions, initiate immediate suppressive fires and take cover, shifting fires as those in the kill zone assault through the ambush (see figure 5-14).

The immediate assault during a near ambush should be a decentralized action. However, it is very important that the squad leader not lose control of the squad after the immediate assault. The squad leader must immediately retake control to assess the situation and decide the next move. The Marines in the squad must pay attention for guidance from their squad leader as the immediate assault culminates or turns into a more deliberate effort.

In a far ambush (i.e., beyond 50 meters or beyond hand grenade range), Marines immediately return fire while taking cover and suppress the enemy. They concentrate on destroying enemy crew-served weapons first, obscuring the enemy position with smoke, and keeping up suppressive fire. Marines not receiving fire move by covered and concealed routes to a vulnerable enemy flank

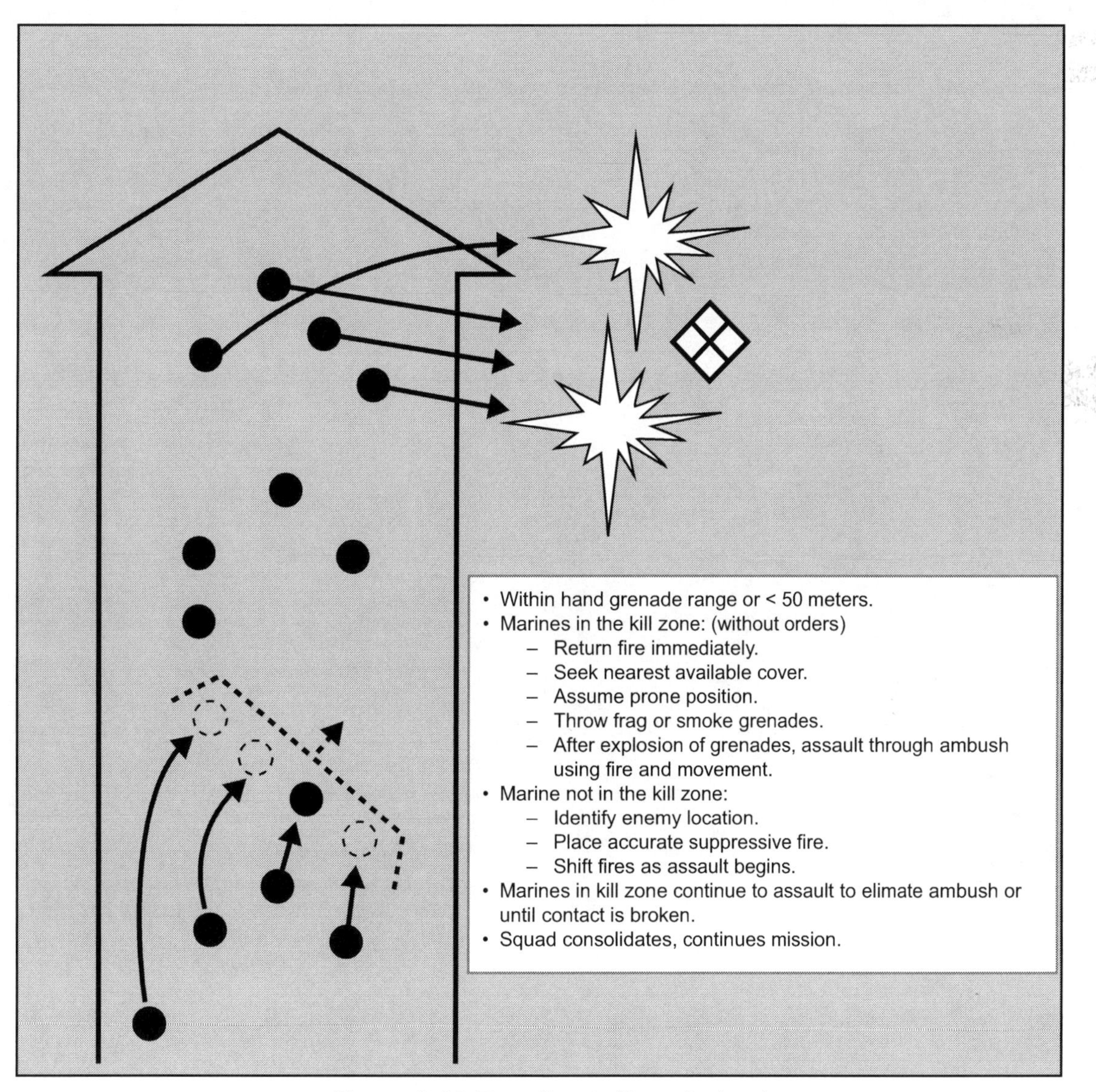

Figure 5-14. Reaction to Near Ambush.

and assault using fire and movement techniques. Marines in the kill zone continue suppressive fire and shift fire as the assaulting element fights through the enemy position (see figure 5-15).

Break Contact. The squad may find itself being engaged by enemy forces by either direct or indirect fire. In either instance, the squad must conduct an immediate action drill to break contact. The two methods to break contact are the clock system and fire and maneuver.The clock system is usually used when the squad is being engaged by enemy indirect fires and the source or direction of fire cannot be identified (see figure 5-16). The method is executed as follows:

- The squad leader shouts or gives the hand-and-arm signal indicating a direction and a distance (e.g., *Ten o'clock; two hundred*).
- Marines move in a ten o'clock direction for 200 meters.
- Patrol members keep the same relative positions as they move.
- Subordinate leaders must ensure the members of their elements and teams move as directed.
- Movement continues until contact is broken.

Note: Twelve o'clock is the direction of movement of the patrol.

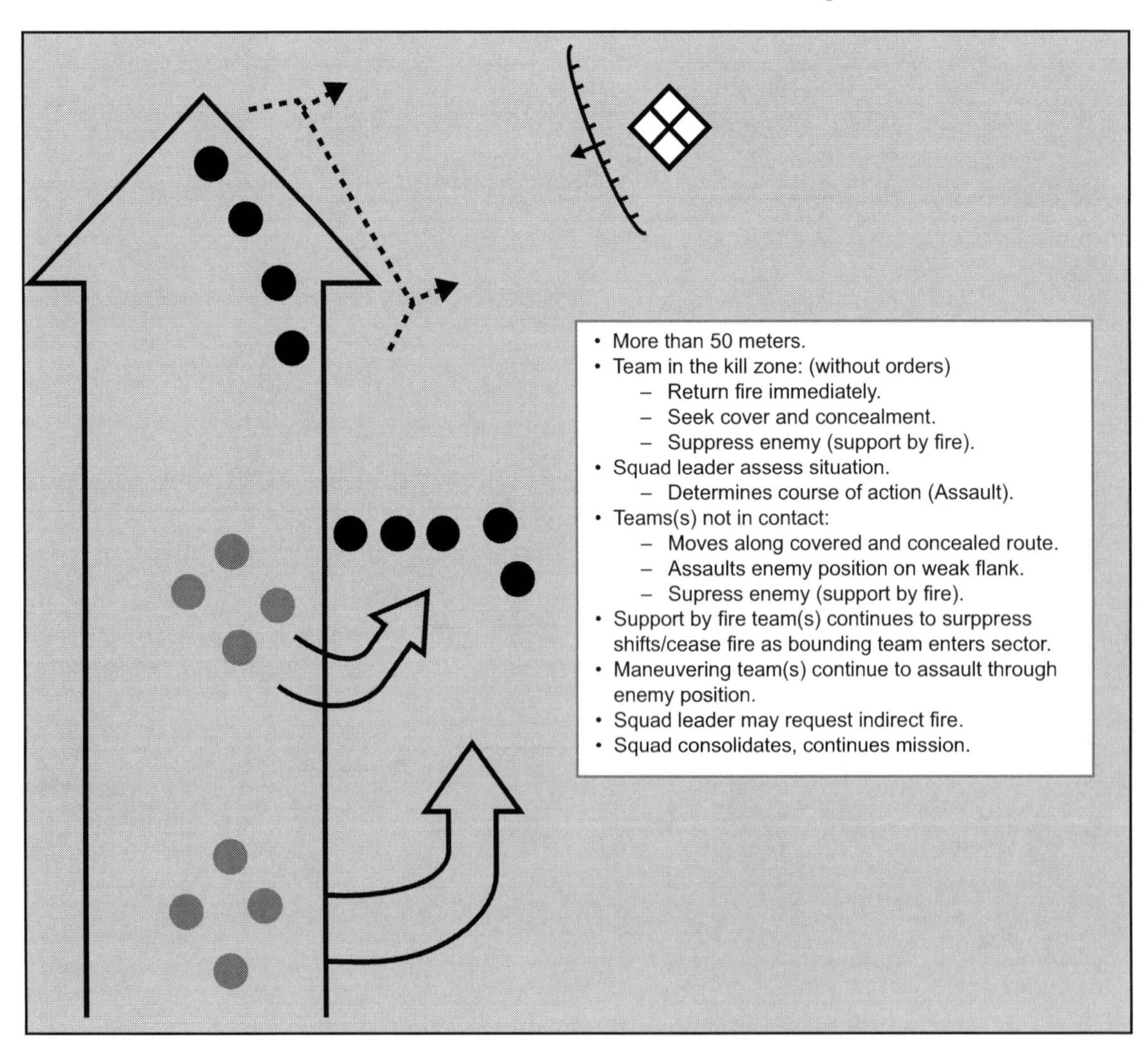

Figure 5-15. Reaction to Far Ambush (assault).

The fire and maneuver method of breaking contact is usually conducted when the patrol is being engaged by enemy direct fire weapons, as in a far ambush, and the squad leader determines breaking contact is required vise assaulting the enemy's position (see figure 5-17). The squad takes the following actions:

- The squad leader directs one fire team in contact to support the disengagement of the remainder of the patrol.
- The squad leader orders a distance and direction, or the last rally point, for the movement of the patrol not suppressing.
- The base of fire continues to suppress the enemy.
- The moving element uses smoke grenades or supporting fires to obscure its movement.
- The moving element takes up designated positions and engages the enemy.
- The squad leader directs the base of fire element to move to their location.
- The squad continues to bound away from the enemy, continuing to suppress until contact is broken.

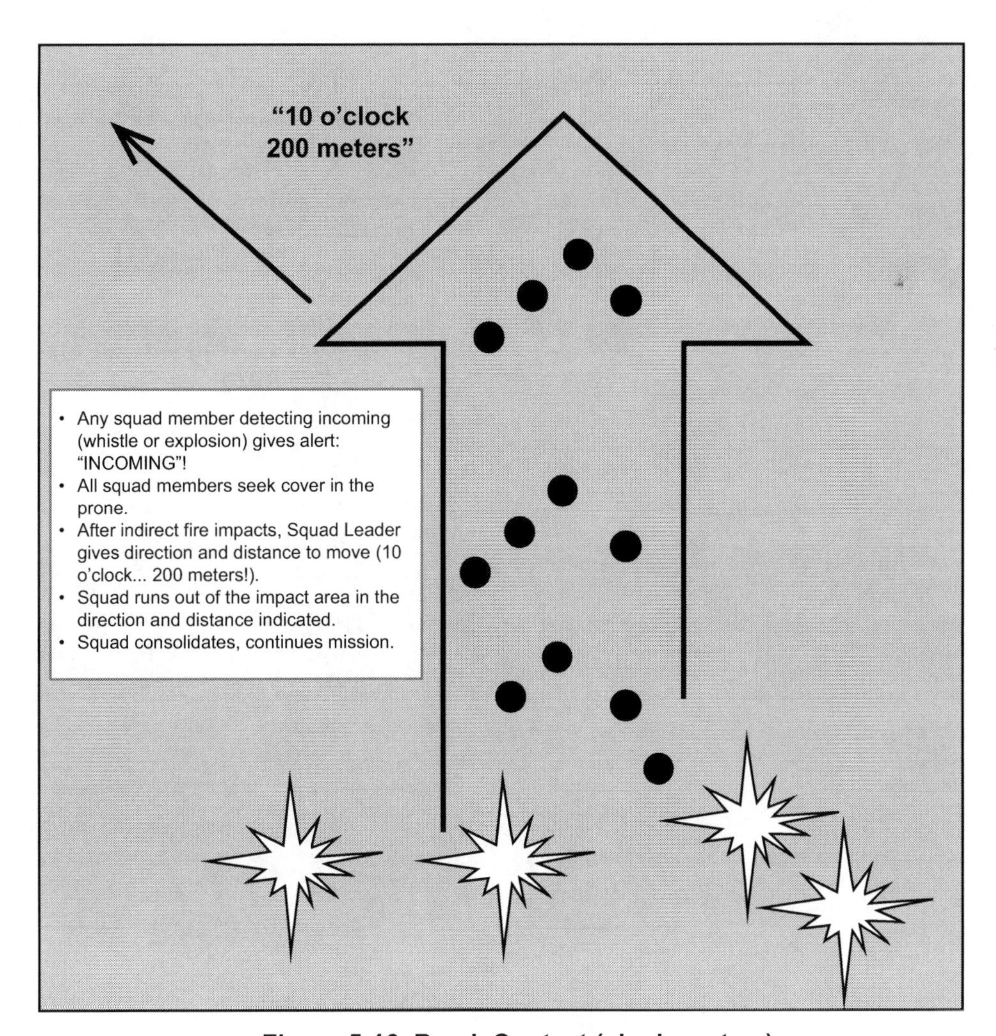

Figure 5-16. Break Contact (clock system).

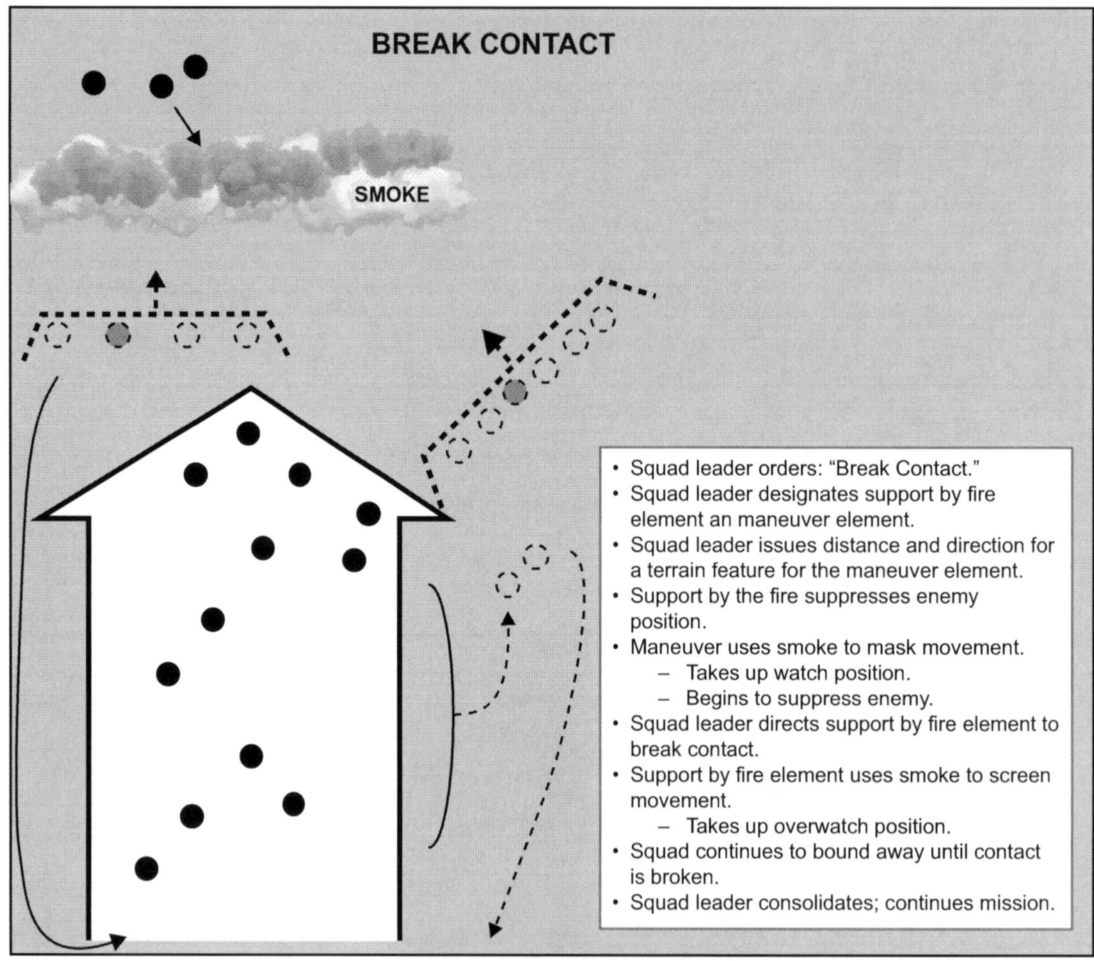

Figure 5-17. Break Contact (fire and maneuver).

Danger Areas

A danger area is any place where the patrol is vulnerable to enemy observation, fire, or both. While patrols typically avoid danger areas, sometimes they cannot. When it must cross a danger area, it does so as quickly and as carefully as possible. The squad leader plans for crossing each known danger area and includes these plans in the patrol order. Table 5-2 shows a list of danger areas and actions to be conducted when encountering them. For more information on danger areas, see MCTP 3-01A. The squad leader decides how the unit will cross based on the following:

- Time available.
- Size of the unit.
- Size of the danger area.
- Fields of fire into the area.
- The amount of security that can be posted.

The squad may cross the danger area all at once, in buddy teams, or one at a time. Regardless of the crossing technique, as each element crosses it moves to an overwatch position or to the far side rally point until told to continue movement.

Table 5-2. Danger Areas and Actions at Danger Areas.

Danger Area	Action at Danger Area
Open Areas	Conceal the squad on the near side and observe the area. Post security to give early warning. Send an element across to clear the far side. When cleared, cross the remainder of the squad at the shortest exposed distance and as quickly as possible.
Roads and Trails	Cross roads or trails at or near a bend, a narrow spot, or on low ground.
Villages	Bypass villages on the downwind side and well away from them. Avoid animals, especially dogs, which might reveal the presence of the platoon.
Enemy Positions	Pass on the downwind side (the enemy might have scout dogs). Be alert for trip wires and warning devices. Indirect fires may be used to divert the enemy's attention.
Explosive Hazards	Explosive hazards can consist of IEDs, mines, and other threats. An individual patrol should avoid deliberate minefields if possible, but a guard for a larger unit may have to address them. Combat engineer attachments are highly valuable if prior planning indicates a high probability of danger areas with explosive hazards.
Streams	Select a narrow spot in the stream that offers concealment on both banks. Observe the far side carefully. Emplace near side and far side security for early warning. Clear the far side and then cross rapidly but quietly.
Wire Obstacles	Avoid wire obstacles whenever possible (the enemy usually covers obstacles with observation and fire).

Patrol Base Operations

A patrol base is a position set up when the squad conducting patrolling activities halts for an extended period. Patrol bases should not typically be occupied for longer than 24 hours except in an emergency. The squad leader selects the tentative site from a map study or by aerial reconnaissance. Plans to establish a patrol base must include selecting an alternate site, which is used if the primary site is unsuitable or if the squad must unexpectedly evacuate the first patrol base. The squad should never use the same patrol base twice. Patrol bases should be used to—

- Stop all movement to avoid detection.
- Hide during a long, detailed reconnaissance of an objective area.
- Eat, clean weapons and equipment, and rest.
- Plan and issue orders.
- Reorganize after infiltrating an enemy area.
- Have a base from which to conduct several consecutive or concurrent operations, such as ambushes, raids, reconnaissance, or security.

Planning Considerations

Squad leaders must consider mission accomplishment and security measures (passive and active) when selecting locations for patrol bases. Security measures to consider include terrain that—

- The enemy might consider of little tactical value (e.g., dense vegetation, swampy, formidable).
- Is off main lines of drift.
- Is near a source of water.
- Can be defended for a short period and that offers good cover and concealment.
- Provides more than one avenue of escape.

The tentative position should be chosen to allow for the following:

- OPs.
- Communication with OPs.
- Defense of the patrol base.
- Withdrawal from the patrol base, including withdrawal routes, rally points, rendezvous points, or an alternate patrol base.
- A security plan to ensure that designated Marines are awake at all times.
- Enforcement of camouflage, noise, and light discipline.
- Required activities are performed with minimal movement and noise.When selecting patrol base locations, the squad leader should avoid—
  - Known or suspected enemy positions.
  - Built-up areas (when not operating in an urban environment).
  - Ridges and hill tops (except as needed for maintaining communication).
  - Roads and trails (i.e., natural lines of drift).
  - Small valleys.

Occupation

A patrol base must be occupied using stealth and deception to maintain secrecy. The patrol base should typically be occupied during periods of limited visibility. Secrecy is maintained by practicing deception techniques that are carefully planned. Deception plans should include the following considerations:

- The route selected avoids centers of population.
- Inhabitants that cannot be avoided are deceived by marching in a direction which indicates that the patrol is moving away from the area.
- Patrol bases are usually located beyond areas that are patrolled daily.
- Typically, not more than one trail (camouflaged and guarded) should lead into the base.
- The base is occupied as quickly and quietly as possible.
- Terrain features that are easily identified are selected as checkpoints.
- If necessary, local inhabitants met by the patrol in remote areas are detained.

Care must be taken if the patrol conducts the occupation under NVDs to limit emitting light signatures; threat forces should always be considered to possess like capabilities and equipment. A squad of three fire teams may occupy a patrol base in the following manner (this may be used by a larger force as well):

- The patrol halts at the last suitable position, approximately 200 meters from the tentative patrol base location.
- The squad leader selects the point to leave the route of march.
- Close-in security for the patrol is established, and the patrol leader and subordinate leaders conduct a leader's reconnaissance (see figure 5-18).
- The patrol leader moves to the tentative patrol base location and designates the point of entry into the patrol base location as 6 o'clock, then moves to and designates the center of the patrol base as patrol headquarters.
- Subordinate leaders reconnoiter areas assigned to them by the clock system for suitability and return to the patrol leader upon completion.
- The patrol leader sends the team leaders to bring the patrol forward (see figure 5-19).
- The patrol leaves the line of march at right angles and enters the base in single file, moving to the center of the base. Designated Marines remove signs of the patrol's movement.

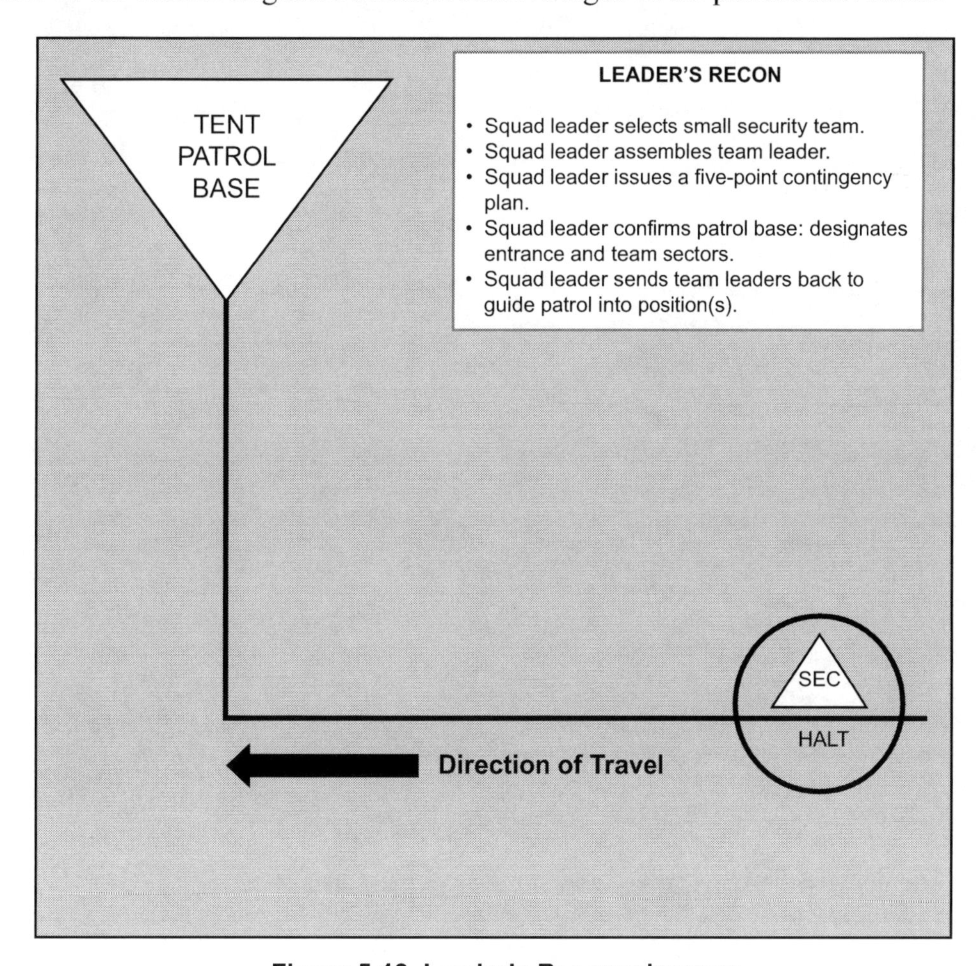

Figure 5-18. Leader's Reconnaissance.

- Each leader peels off their unit and leads it to the left flank of their assigned sector.
- Each unit occupies its portion of the perimeter by moving clockwise to the left flank of the next sector (see figure 5-20).
- Each unit will then stop, look, listen, and smell—known by the acronym, SLLS—before reconnoitering forward of its sector.
- Each unit then reconnoiters forward of its sector (this should be METT-T dependent during limited visibility) by having designated individuals move a specified distance out from the left flank of the sector, moving clockwise to the right limit of the sector, and re-entering at the right flank of the sector. They report indications of enemy or civilians, suitable LP/OP positions, rally points, and withdrawal routes.
- The patrol leader then designates rally points, positions for LP/OPs, and withdrawal routes.
- Each fire team then puts out a LP/OP in front of each sector, establishes communications, and commences base routine.

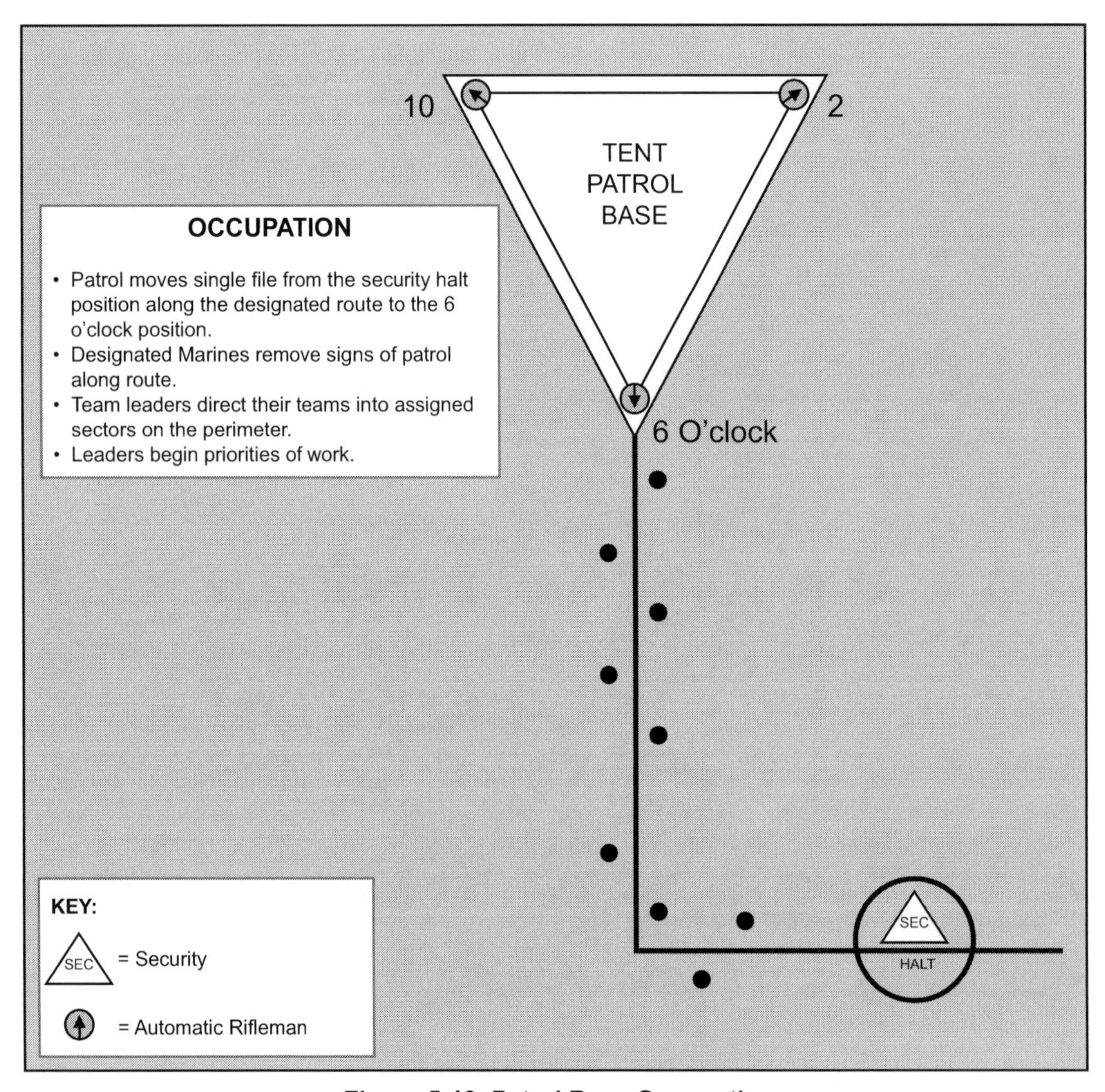

Figure 5-19. Patrol Base Occupation.

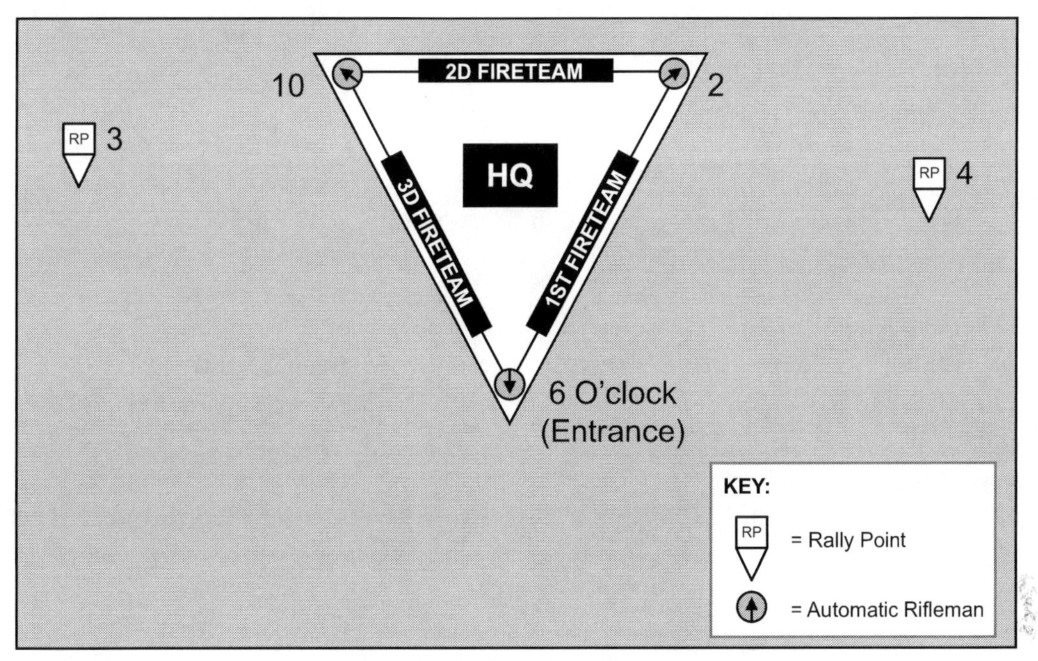

Figure 5-20. Occupied Patrol Base.

Base Routine

Once the squad leader determines the area is suitable for a patrol base, they establish or modify defensive work priorities in order to establish the defense for the patrol base. Priorities of work are not a laundry list of tasks to be completed in sequence. The squad leader must prioritize these activities based on METT-T. For each priority of work, a clear standard must be issued to guide Marines in the successful accomplishment of each task. Priorities of work may include, but are not limited to, the following:

Alert/Stand-To. The squad leader announces the alert posture and the stand-to time. Plans are developed to ensure all positions are checked periodically, that OPs are relieved periodically, and that at least one leader is always alert. The patrol typically conducts stand-to at a time specified by unit SOP, such as 30 minutes before sunrise or sunset. At stand-to, all gear and equipment is packed and prepared for immediate movement, should the situation arise.

Resupply. Distribute or cross-load ammunition, meals, equipment, etc.

Feed Plan. At a minimum, security is established and weapons maintenance performed prior to eating. Typically, no more than half the squad should eat at one time. Marines should typically eat one to three meters behind their fighting positions.

Water Resupply. The squad leader organizes watering parties as necessary. The watering party carries canteens and water bladders in an empty rucksack or other bag, and must have communications and a contingency plan prior to departure. Marines must be equipped and prepared to utilize water purification methods (e.g., tablets, filters) when sufficient unit water resupply is unavailable.

Weapons and Equipment Maintenance. The squad leader ensures that individual and crew-served weapons, communications equipment, NVDs, and other special equipment are maintained. No more than one-third of all weapons are disassembled at a time and weapons are not disassembled at night.

> *Note:* If one machine gun or automatic rifle is down, then security for all remaining systems is raised.

Sanitation and Personal Hygiene. The squad leader and corpsman ensure a slit trench (if applicable) is prepared and marked. The patrol must not leave trash behind.

Departing the Patrol Base

Before leaving the base, all signs of occupation are removed. The area is left to appear as though it had not been occupied. The squad typically departs the patrol base in the reverse order from how it was occupied.

CHAPTER 6
MILITARY OPERATIONS ON URBANIZED TERRAIN

NATURE OF MILITARY OPERATIONS ON URBANIZED TERRAIN

Military operations on urbanized terrain is a broad term referring to the many actions and considerations necessary in applying the fundamentals of warfare to the complex urban environment. The basic tenets of rifle squad employment do not change because the squad enters urban terrain, and the rifle squad must train and adapt to urban terrain the same way it would to any other environment. This chapter should be viewed as a quick reference guide for tactical actions and planning considerations. For more information on military operations on urbanized terrain, see MCRP 12-10B.1, *Military Operations on Urbanized Terrain (MOUT).*

Environment

As civilization progresses, militaries can expect to find themselves fighting for control of urban areas. Unconventional forces understand that to face a conventional force in the open country would be equivalent to suicide. Additionally, forces may find embedding in the obstacle-strewn and complex urban environment provides significant advantages in cover and concealment while also positioning them to control large civilian population and resource bases.

Urban areas vary widely in their makeup and classification. Trends in urbanization have led to the construction of mega-cities housing millions of civilians alongside government and military infrastructure. These cities may also include urban slums, compact and often ad hoc groupings of low income people in low quality and often overcrowded housing. Squads may also find themselves being deployed to less-developed urban areas, such as traditional rural villages with low levels of infrastructure, and family housing often separated by high compound walls.

One of the core challenges of urban operations is the battlefield geometry. Rather than the largely two-dimensional battlefield of open areas, urban areas require Marines to clear their sectors of fire vertically, as well as left and right.

Another fundamental challenge of operating in urban environments is the problem of massing combat power. Urban areas often cause reduced dispersion as compared to open areas. This is due to the close-range nature of the threat, as well as the complex and numerous avenues of approach and sectors of fire that need to be covered. Whereas a rifle squad may be able to cover a few hundred meters of temperate forest, in the urban environment a squad may only be able to clear and hold one floor of a complex structure.

In addition to the vertical element introduced by buildings, walls, and other urban terrain, there may often be a subterranean component of the urban environment that cannot be ignored. Underground passages such as sewage lines or tunnel systems may offer the adversary additional lines of communication and increased mobility and vectors of attack. Threats can bypass urban defenses through subterranean networks without being noticed, and tunnels may be used to escape cordon and search operations.

Varying Intensity

Since urban operations occur in areas heavily populated by civilians, squad leaders must be aware of the diverse requirements for successfully operating in a fluid environment. It is entirely possible that Marines in one city block could be involved in low intensity civil-military operations such as humanitarian aid distribution, possibly partnered with host nation forces, while Marines in the next city block are required to execute security and police actions such as vehicle control point operations, and Marines in the next city block are required to engage in combat operations such as breaking an enemy ambush and destroying the enemy threat. The squad leader must have the mental flexibility to adjust their Marines' mindset up and down the scale of intensity; often in the same day. This requires intense military discipline and exhaustive contingency planning.

Initiative-Based Tactics

Initiative-based tactics is a core enabling requirement of urban combat. Due to the complexities of the battlefield and the fluid operational environment in the urban setting, the traditional chain of command is often inadequate for rapidly addressing the challenges faced at the tactical level. As such, through training and rehearsal, each Marine in the rifle squad must be empowered to make tactical decisions in response to the changing and evolving environment and threat. The following principles must be accounted for in shaping the decisions of individual Marines, and will often be the difference between success and failure in rapid and complex tactical urban operations.

Principles

Security. The myriad potential for three-dimensional geometries of fire in urban operations precludes the squad leader from accurately predicting all possible threats. Individual Marines must maintain high levels of situational awareness regarding the threat, the friendly force intent, and the operational environment, and take the initiative in covering fields of fire or adjusting positions to best mitigate the security threat to friendly forces. This is a continuing action, as both leadership and individual Marines must be constantly appraised of the effectiveness of security. Marines must be empowered to adjust their courses of action to account for emerging threats and the complexities of the urban environment.

Surprise. Marines may utilize surprise to gain the advantage over an enemy through well executed stealth and tempo. Surprise may also be leveraged to take advantage of chance events on the battlefield, such as discovering previously unknown covered routes to an objective or identifying an unexpected enemy high valued target. Despite best laid plans, Marines must be trained to recognize when surprise may be utilized to generate more desirable outcomes within the commander's intent.

Tempo. Speed is critical in urban operations, but not for the sake of speed itself. Rather than thinking in terms of speed, Marines should frame their actions in terms of comparative tempo between themselves and the enemy. Executing smooth, methodical, and effective tactical action at a tempo that exploits battlefield events faster than the enemy is more effective than risking effectiveness and security to beat an imagined clock. Marines should always remember the saying, "slow is smooth; smooth is fast."

Accuracy. Due to the compartmentalized nature of urban terrain, it is easy for Marines to become disconnected from battlefield events, regardless of their relative proximity. It is the responsibility of every Marine to ensure that a shared situational awareness is maintained throughout tactical actions. Clear, concise commands and reports should be utilized early and often.

Ad Hoc Organization

As discussed in the Interior Movement portion of this chapter, it is easy to see how the traditional organization of the rifle squad may become shuffled during tactical operations. While the chain of command will always hold ultimate authority, individual Marines must be empowered to make the tactical decisions required by their physical position in the environment, not by their rank alone. For instance, as a squad clears through a structure, clearing it room by room, the squad leader may become separated from the lead element of the squad, while the most junior private first class becomes the number one Marine (i.e., the first Marine in the stack to enter a room or structure). To operate effectively, even that private first class must be empowered to continue clearing, call for support, or take other tactical action in the absence of direct orders based on the Marine's own judgment and initiative.

Violence of Action. Urban combat is often fast-paced, confusing, and frantic. The principle violence of action means that Marines engaged in urban combat must take the necessary actions to maintain the initiative once in contact. Accurate fire, aggressive movement and maneuver, and shock effects such as explosive breaching and clearing should be utilized to establish and maintain dominance on the urban battlefield.

These principles can only be effectively employed through strict discipline and constant training and rehearsal. The squad leader must impress the correct tactical mindset on all the members of the squad, and—in turn—it is crucial that each member of the squad become familiar with each other's thought processes and begins to trust in the initiative-based tactics their fellow Marines are likely to employ. At minimum, these principles of initiative-based tactics should drive Marines to adhere to three core rules: identify and cover danger areas, accurately engage threats, and protect each other.

Identify and Cover Danger Areas. Any area that has not been cleared or cannot be confidently determined to be clear of threats is considered a danger area. This includes areas previously cleared but left unoccupied. Marines must be able to identify danger areas as they arise and ensure they are adequately covered. Marines should not give up covering their danger areas until relieved by other Marines, or until the unit moves beyond the potential impacts of the danger area.

Accurately Engage Threats. The importance of differentiating between threat and friendly forces or civilians must be in the forefront of every Marine's mind before employing any weapon. Once threats have been accurately identified, Marines can then accurately and purposefully employ their available firepower.

Note: Accurate fire is effective fire.

Knowing the target, what lies beyond it, and what is between the weapon and the target is crucial in an urban environment, where rounds can continue to penetrate through walls or become deflected by other objects.

Protect Each Other. Marines must take the initiative to identify gaps in security, communicate threats, and cover the actions of their fellow Marines and/or partner forces. Security is the responsibility of the whole unit, and it is only accomplished through mentally engaged individual Marines taking the initiative. Communications must be continuous to keep up the whole unit's situational awareness and empower more initiative-based decisions.

EXTERIOR MOVEMENT

The fundamental principles of dispersion, fore and aft positions, mutual support, and overwatch continue to apply in the urban environment. Squad leaders execute exterior movement in accordance with the threat environment and tactical tempo requirements. When selecting routes of movement, the squad leader should recall that the urban environment offers opportunities for nonstandard movement, such as—

- Utilization of subterranean access points.
- Traversing through buildings (e.g., breaching walls and barriers).
- Utilizing ladders to gain access to higher levels or to mitigate walls.

These should be considered when selecting routes rather than simply sticking to established roads and paths. Creativity tempered with practicality in route selection may make the difference in gaining the initiative on the battlefield and foiling the enemy's expected courses of action. For more information on urban movement, see MCRP 12-10B.1.

Danger Areas

Danger areas can be crossed utilizing the bump or bound methods. The fundamental requirement for each of these techniques is the need for persistent security covering the danger area. Examples of both methods are depicted in figures 6-1 through 6-6.

Bump. Figure 6-1 shows the number one Marine providing initial security for the danger area.

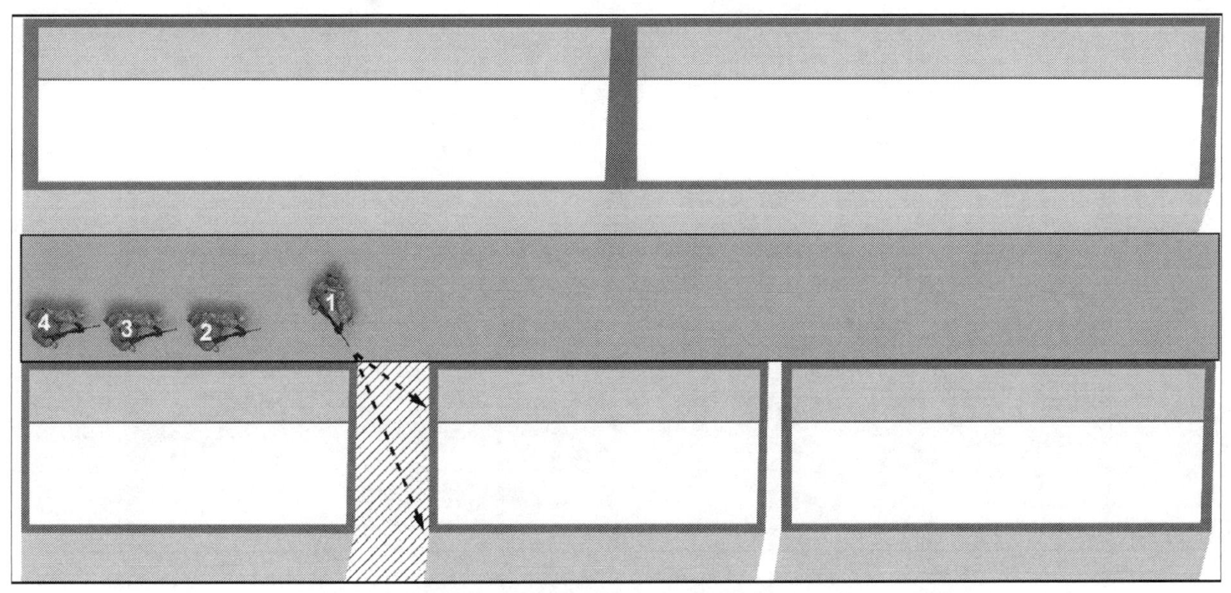

Figure 6-1. Bump Method (providing initial security).

Figure 6-2 shows the number two Marine assuming security while "bumping" the number one Marine to continue on the route.

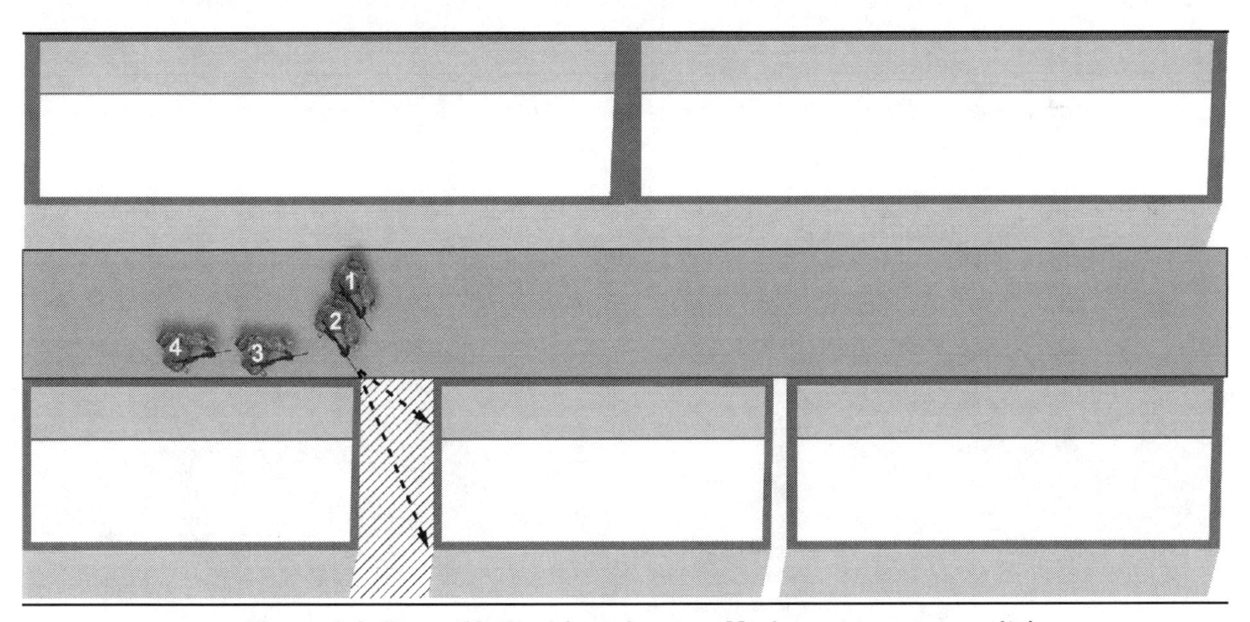

Figure 6-2. Bump Method (number two Marine assumes security).

Figure 6-3 shows the remaining members assuming security and "bumping" until they clear the danger area.

Figure 6-3. Bump Method (Marines continue to "bump").

Bound. This method differs from the bump method, in that after the number one Marine identifies the danger area (see figure 6-4) and establishes security, the remaining team members all move past the number one Marine and continue on the route (see figure 6-5). Once they have passed the number one Marine, that Marine assumes the position of rear security for the team (see figure 6-6).

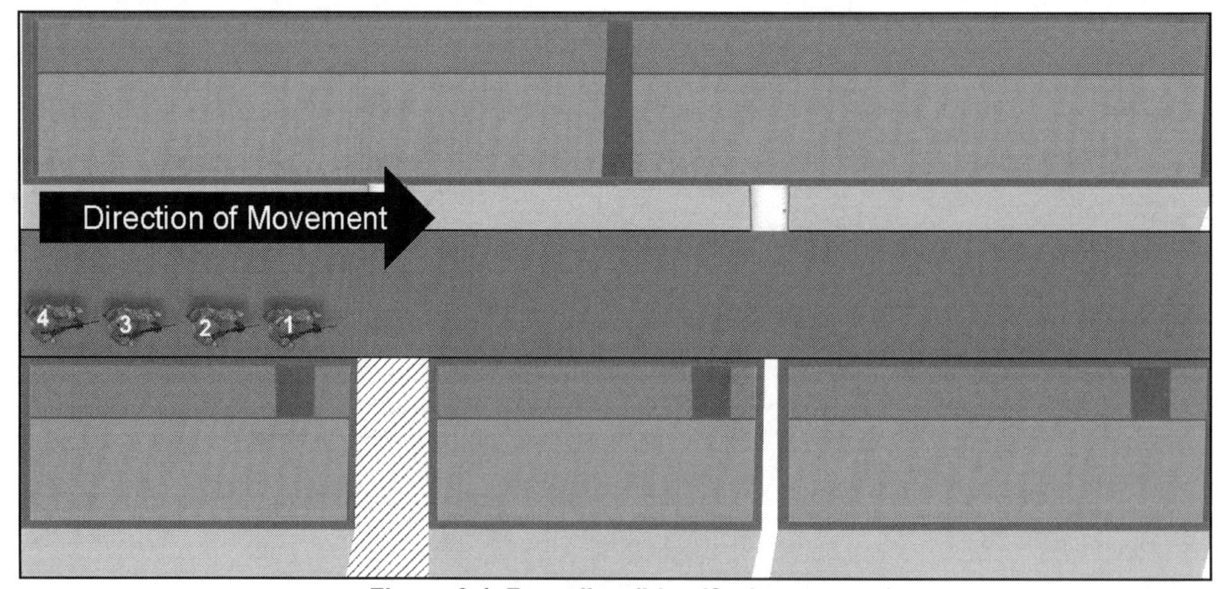

Figure 6-4. Bounding (identify danger area).

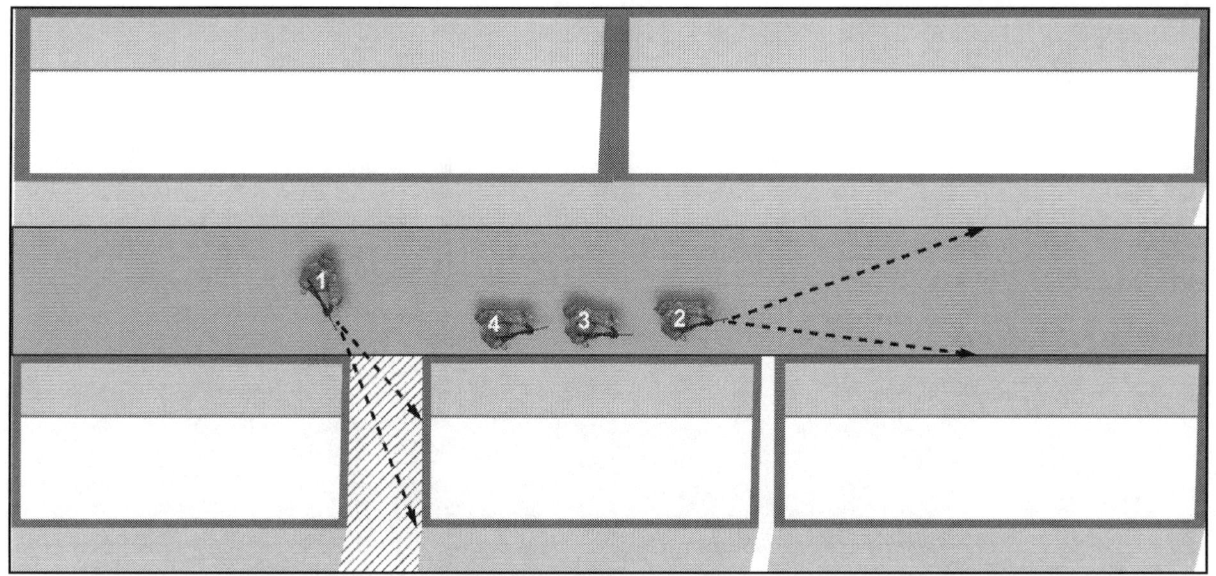

Figure 6-5. Bounding (establish security, team "bounds" past).

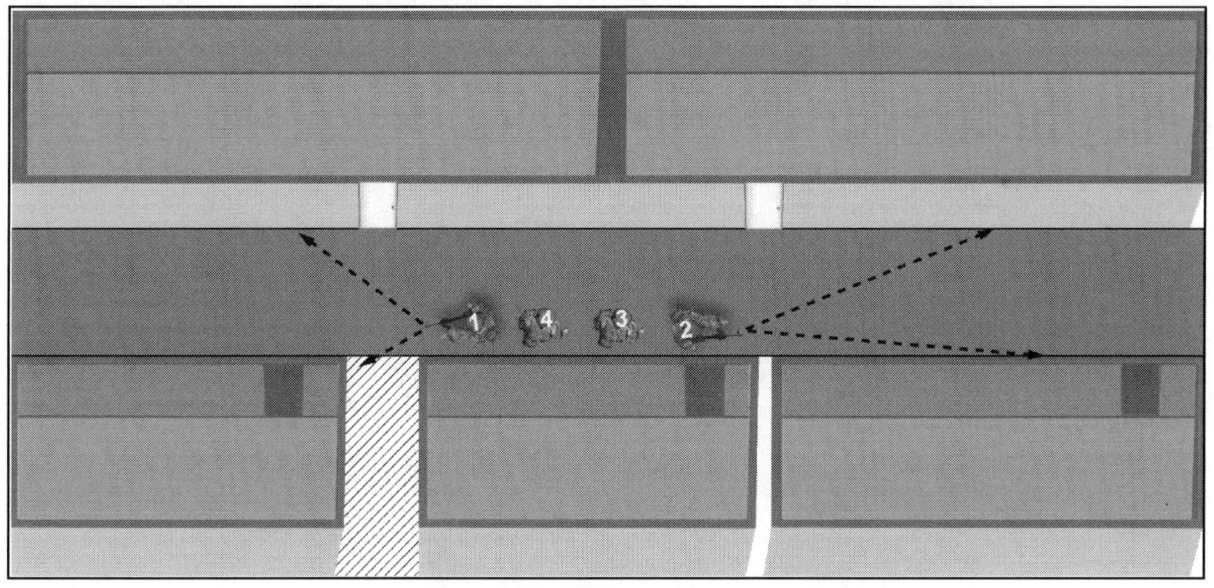

Figure 6-6. Bounding (assuming rear security).

Double Column. Double column is a method of movement that offers greater mutually supporting cover from close-range threats. It is generally used when threats are expected from above ground floor level. In a double column, the squad divides itself between both sides of a street or other avenue of approach.

Marines at the front and rear of the column cover their respective fields of fire, while the remaining Marines provide mutual security over their opposite members across the avenue of approach. For instance, the Marines on the left side of a street would cover the upper level windows or danger areas above the Marines on the right side of the street and vice versa (see figure 6-7).

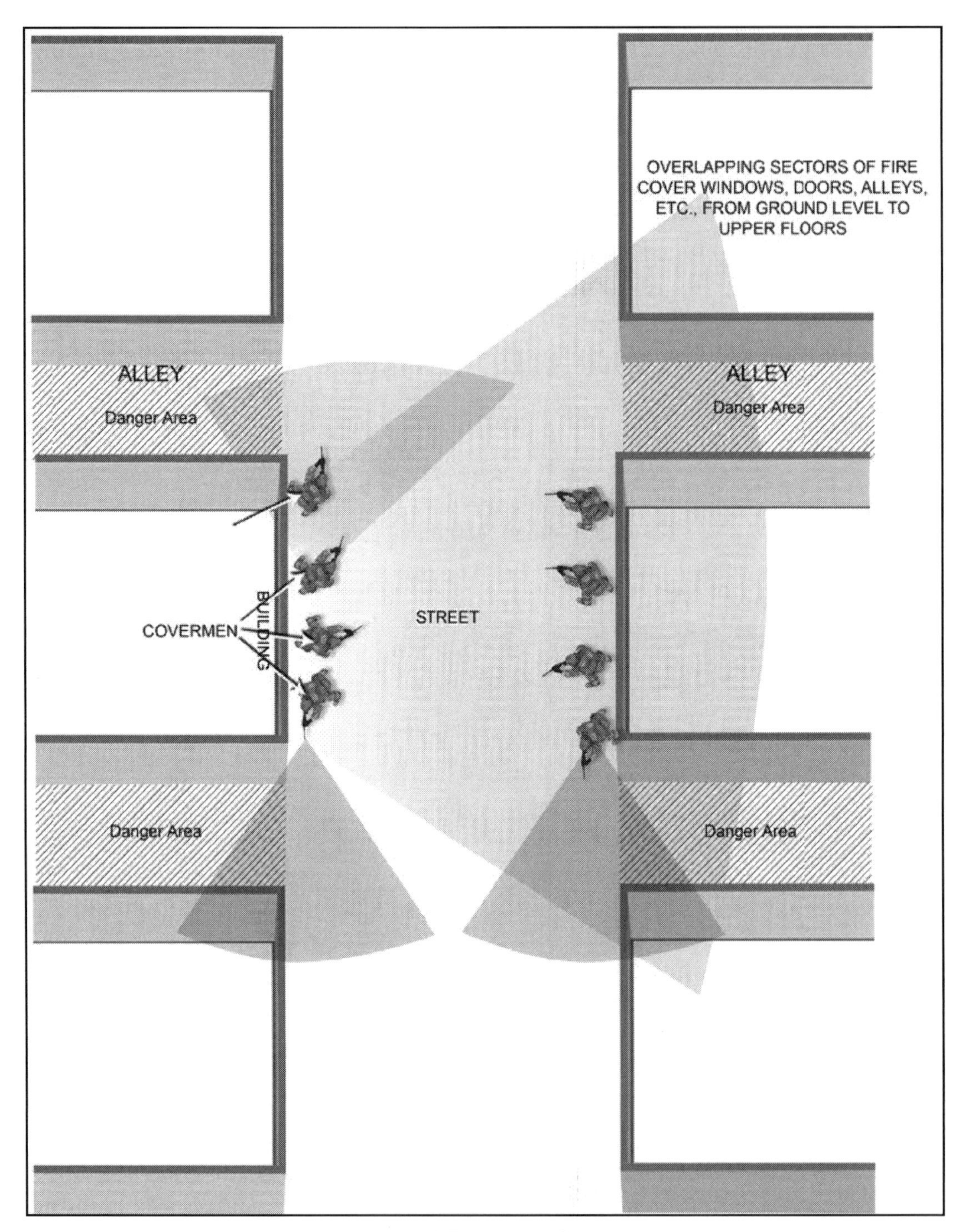

Figure 6-7. Double Column.

Satellite Patrolling

In a semi-permissive or unknown threat environment, the satellite patrolling method may be utilized to confuse threat forces, maximize mutual support, and establish a greater depth of security for the squad. Satellite patrolling utilizes one or more "satellite" elements that move semi-independently of the main patrol's direction of march. These satellites follow the general direction of march, but are free to patrol in a more ad hoc fashion while remaining within mutual supporting range of the main body of the patrol. This expands the patrol's security envelope while also serving to confuse the enemy as to the patrol's true location, intentions, and direction of march. Satellite patrolling is executed utilizing one of the following methods: zig-zag, swinging the gate, satellite (revolutions), double back, or a combination (see figure 6-8, figure 6-9, figure 6-10, and figure 6-11).

Satellite patrolling requires the squad to maintain positive implicit and explicit communications. The squad should also rehearse actions on contact for each satellite and for the main body to ensure the squad can react to a variety of potential engagement scenarios.

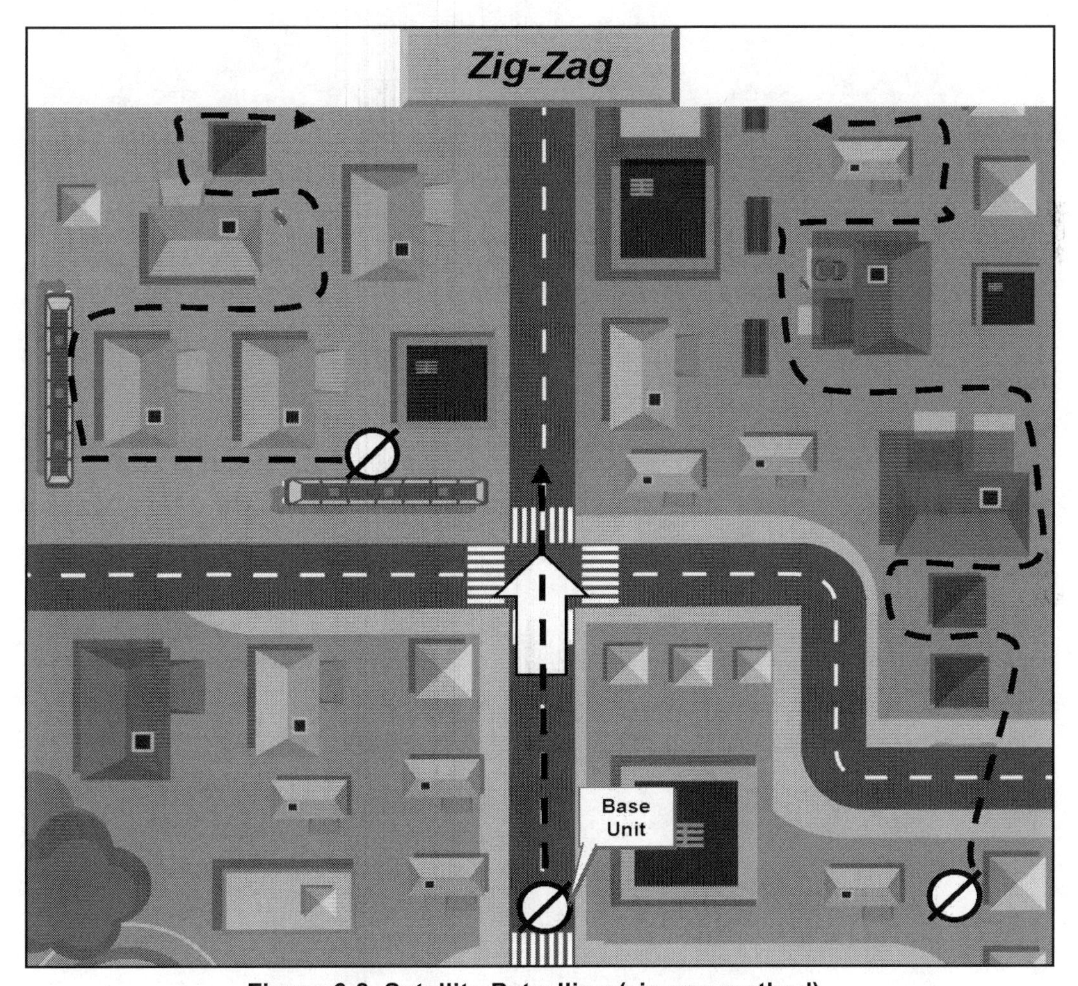

Figure 6-8. Satellite Patrolling (zig-zag method).

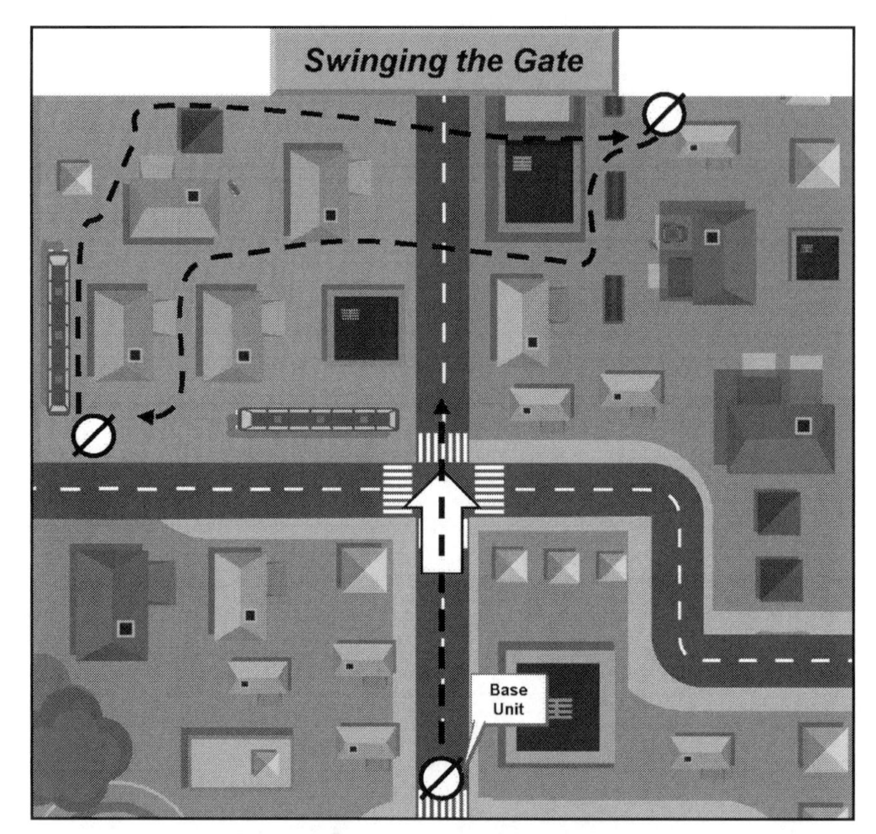

Figure 6-9. Satellite Patrolling (swinging the gate method).

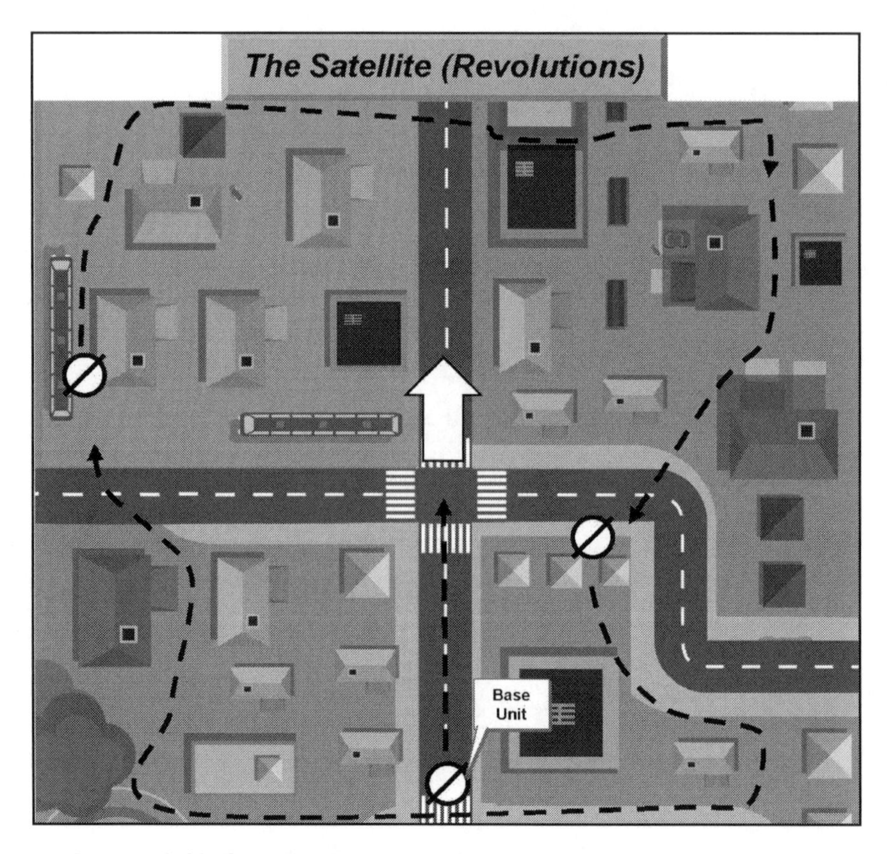

Figure 6-10. Satellite Patrolling (satellite revolutions method).

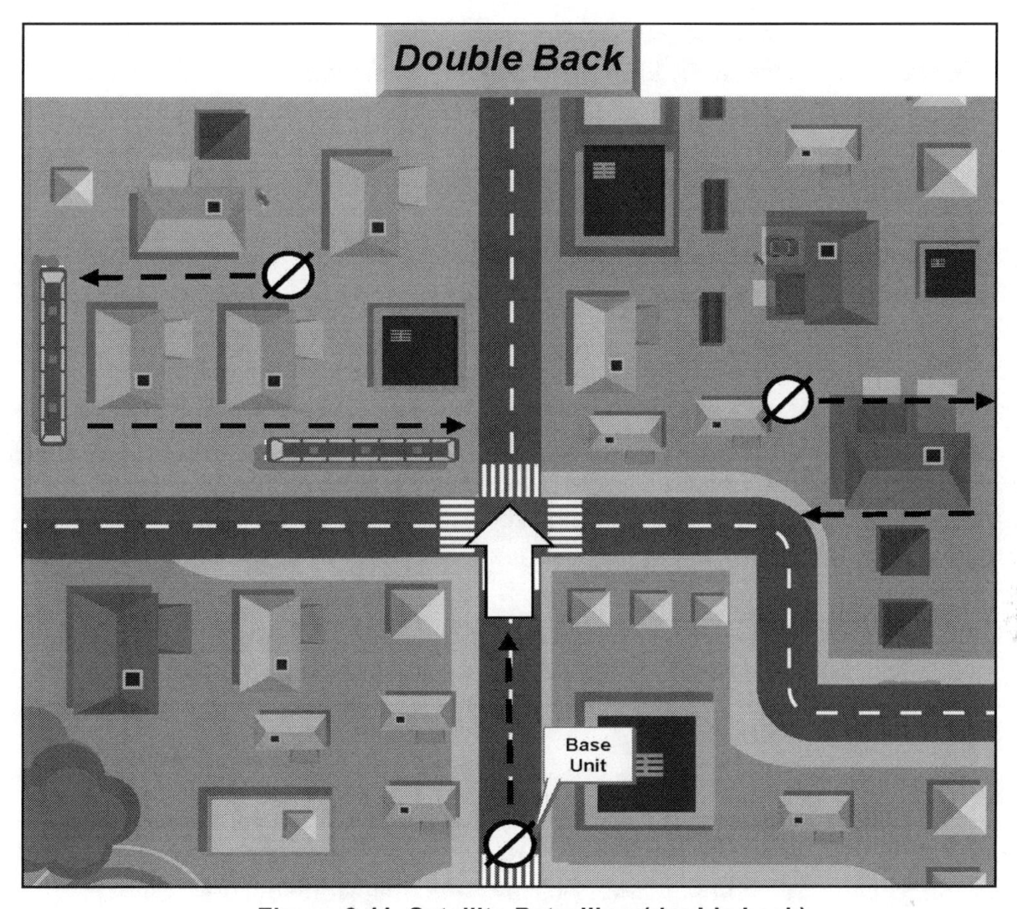

Figure 6-11. Satellite Patrolling (double back).

Cordon and Search

A cordon and search may be conducted in any threat environment, from permissive to hostile. This is a technique for isolating a target area from outside influence, preventing escape from the target area, and conducting a thorough search of the target for any of a variety of reasons. The rifle squad is sufficiently equipped to cordon and search small structures or small compartmentalized urban terrain features. However, due to manpower requirements, the rifle squad can often find itself working with the platoon or company to accomplish larger and more complex cordon and searches. During these operations, the squad may be assigned to one of three main task organized elements: the outer cordon, the inner cordon, or the search team.

Outer Cordon. This element prevents outside influences from interfering with the objective area, whether the influences are civilian traffic or enemy units. The outer cordon allows the inner cordon and search elements to complete their assigned tasks by mitigating outside threats and controlling access to the search area (see figure 6-12).

Inner Cordon. This element is responsible for local security around the objective area and for preventing any personnel or vehicles from departing from the objective area until the search is complete. They are prepared to act as the support element for the search element should the need arise (see figure 6-12).

Search Element. This element conducts the thorough search of the objective site. The search element does not begin their actions until the inner and outer cordons are established. They are responsible for collecting and cataloging the materiel and personnel found in the objective area. The search element leader deliberately plans and communicates a detailed search plan to ensure there are no oversights on the objective. Should the search turn into a hostile engagement, this element transitions roles and serves as an assault element until the threat is neutralized.

A cordon is not a simple, 360-degree perimeter around a target. A 360-degree perimeter would require more Marines than is necessary, and would not necessarily take advantage of available cover and concealment. Rather, Marines pick security positions where there is effective 360-degree coverage on the target while isolating it from outside the cordon by covering terrain with fires and observation. In the example in figure 6-12, due to the geometry of the target building, the rifle squad is able to establish good all-around security by posting fire teams on opposite corners of the compound. The placement of the cordon should be deliberately planned to ensure Marines maintain the most secure and efficient inner and outer cordons possible.

Once all actions on the objective are complete, Marines usually depart the objective in the reverse order of occupation—search/assault element, inner cordon, and outer cordon.

Ladder Employment

One of the most useful pieces of equipment to the rifle squad in the urban environment is the simple ladder. Although it may be ungainly to physically carry, the ladder gives the squad maneuver and mobility options in the vertical direction of the urban environment. The ladder must be rated for the weight of a combat loaded Marine and collapsible to a portable size.

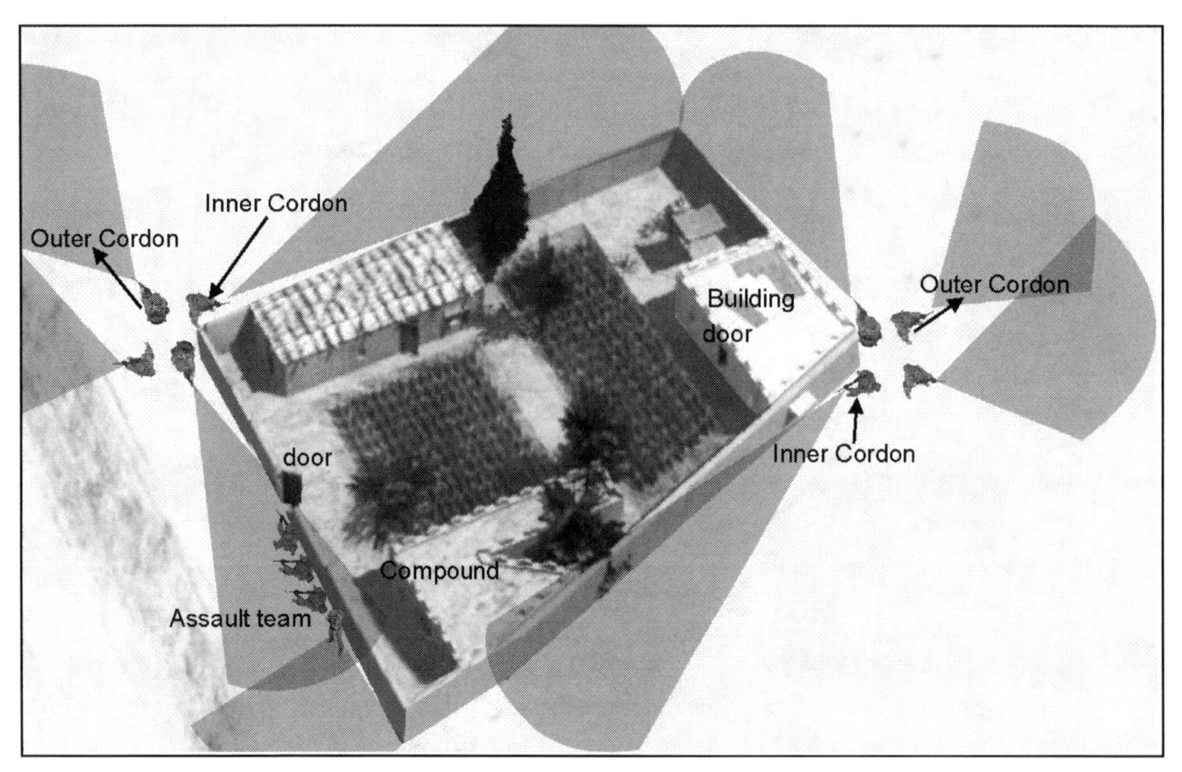

Figure 6-12. Inner and Outer Cordons.

Ladders can be used to gain access to upper levels of an urban area to occupy an overwatch position, secure an entry point into a structure, or traverse walls and other urban obstacles.

The rifle squad may also use ladders to augment the inner cordon in situations where the objective building is surrounded by a compound wall or other similar obstacle (see figure 6-13). In this situation, Marines determine where the most likely courtyard dead space is and position ladders so that Marines are able to clear those spaces prior to the assault element's entry. Marines should take care to synchronize this action and deconflict geometries of fires so that Marines clear the dead space simultaneously and without risk of fratricide.

The urban environment often canalizes a squad's movement and limits it to moving through established pedestrian pathways. Enemies often use this natural obstacle plan in an opportunistic manner to employ IEDs and ambushes against the canalized squad. Ladders can be used in a method known as "steeple chasing," where the squad makes its own path through and over existing structures and obstacles (see figure 6-14). Using this method requires a solid understanding of the route and likely threats, as well as potential three-dimensional geometries of fire when the squad is on various vertical levels at the same time. The method is best controlled by employing a bounding overwatch form of maneuver.

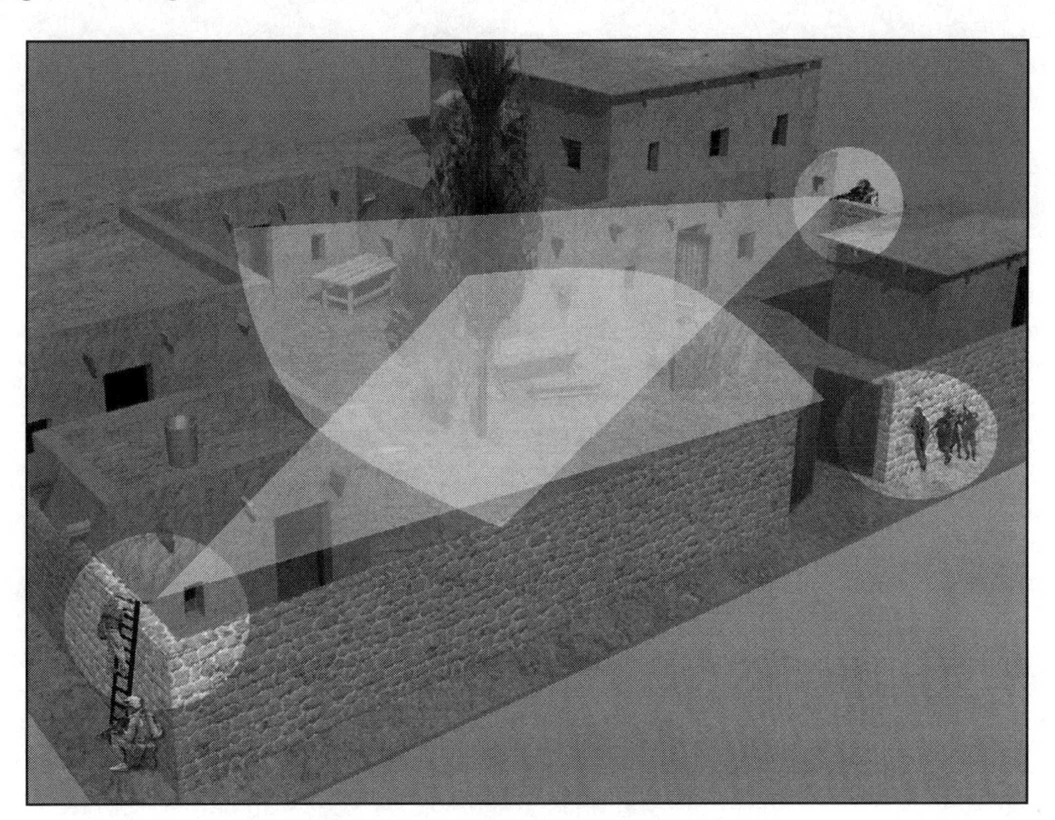

Figure 6-13. Ladder Employment.

Figure 6-14. Steeple Chasing.

Overwatch in the Urban Environment

As with traditional maneuver, utilizing overwatch in the urban environment can add an additional layer of security to the squad. Urban overwatch typically entails securing elevated vantage points from which the overwatch element can observe both friendly forces and the environment beyond the line of sight of forces on the ground (see figure 6-15). This is particularly effective when friendly forces are static and engaged in complicated tasks such as checkpoint operations. However, when static, care must be taken not to expose squad elements to observation and fires by enemy manned or unmanned aircraft, including small UASs. Rather than occupying rooftops, overwatch positions that provide cover and concealment may be more suitable, such as upper-level floors below the roof-top level. In this case, overwatch elements provide all-around security from external threats to friendly forces.

When possible, overwatch positions should be employed in a covert manner to protect the overwatch element. The effective employment of a designated marksman or sniper team in structures (see figure 6-16) is an example of a covert position. The marksman use only the minimum opening necessary to provide a useful field of fire and effective employment of optics (i.e., also known as a loop hole).

Figure 6-15. Overwatch Position.

Figure 6-16. Covert Overwatch Position.

Reconnoiter, Isolate, Gain a Foothold, Seize, Consolidate, and Reorganize Cycle

To transition from exterior to interior movement or to otherwise gain entry to a hostile or potentially hostile emplacement, Marines use the urban attack cycle—reconnoiter, isolate, gain a foothold, seize the objective, and consolidate and reorganize (see figure 6-17). The cycle begins at a covered position, much like an attack position.

Urban Attack Cycle
R – Reconnoiter
I – Isolate
G – Gain a Foothold
S – Seize the Objective
C – Consolidate and Reorganize

Figure 6-17. Urban Attack Cycle.

The urban attack cycle is also used on a smaller scale for the individual clearing actions within a structure or other smaller urban terrain. The cycle can be performed on a large scale such as a battalion-level mechanized assault on an urban phase line, or it can be done on the buddy team level as Marines clear rooms.

Reconnoiter

Prior to making entry, Marines use whatever means are at their disposal to gain more information on their target. This may range from small unmanned systems (i.e., either aerial or ground) to a cursory map and visual estimation. Marines must identify their preferred and alternate points of entry to the target area prior to continuing the urban attack cycle.

Isolate

Marines use various methods to isolate the target area from outside influence to set the conditions for seizing an entry point. In more hostile environments, this is generally accomplished by posting security to cover the flanks of the intended route of movement to the objective and deploying obscuration to mask friendly movement. In more permissive environments, this may be establishing an outer cordon around the objective with a security element.

Gain a Foothold

Marines aggressively and violently dominate the entry point to the objective and penetrate any defensive obstacles using whatever means available—through explosive or mechanical breaching or close combat—to secure the initial foothold in the objective area. In more permissive environments, this means Marines should have absolute positive control over the entry point to the objective. The first cleared area in the objective generally becomes the initial rally point for casualties, reinforcements, and captured personnel and materiel to flow through. Once this entry point is gained, it is relinquished until directed or the unit is relieved by follow-on forces.

Seize the Objective

Gaining a foothold is often the most violent and frantic portion of the urban attack cycle. Once a foothold is established, Marines should methodically clear the objective.

Consolidate and Reorganize

Though included together, consolidation and reorganization usually happen as two separate stages. This phase begins with the immediate actions to repel potential counterattack and includes consolidating the objective area and reorganizing for follow-on tasks. By setting limited objectives, the attacking force has an opportunity to reorganize and defend against counterattack while maintaining momentum. In addition to defensive preparations against counterattack, squad leaders make preparations for follow-on missions (e.g., restoration of civilian facilities, marking and clearing minefields/IEDs). Additional tasks should include—

- Redistributing ammunition and water as soon as possible.
- Evacuating military, civilian, and enemy dead and wounded.
- Organizing detainees for evacuation.
- Resupply to sustain follow-on operations.
- Establishing mutually supporting, in depth, 360-degree defensive positions.

INTERIOR MOVEMENT

Individual Movement

When moving within a structure, Marines should assume an aggressive, mobile stance with their hips, feet, and torso squared to the target while leaning forward to control recoil. In this position, the rifle is carried in a low ready position, eyes over the sights, taking care not to obscure vision with the rifle itself. The Marines square their bodies in relation to the direction of advance in order to maximize the protection offered by their body armor (see figure 6-18). The Marines move in steady, measured steps, transitioning weight from heel to toe, and maintaining a stable shooting platform at all times. When movement has stopped for more than a few seconds, Marines assume the kneeling position and establish local security.

Figure 6-18. Mobile Standing Position.

Commands

Communication within the squad must be rapid and efficient. The following is an example list of common commands along with suggested brevity code words and the actions to be executed:

- Stack. Refers to a team or unit ready to make entry. A stack can be made up of any number of Marines. This term is also used as a verb to direct a number of Marines to a specific location within the objective. After initial entry has been made and a foothold has been established, "stack" will be announced to bring the remainder of the assault element into the objective. During the assault, it may be necessary to direct an element to a specific location within the objective (e.g., *Team one, stack on me*).
  - Noun: A group of Marines ready to make entry (e.g., *The stack is at the door*).
  - Verb: To direct Marines to a specific location (e.g., after initial entry, the call, *Stack left* directs the remainder of the assault element to enter and move to the left).
- Support. This command is used to draw attention to members of the assault element who need assistance. When used alone (e.g., *Support*), only the support of one additional Marine is necessary. If more than one Marine's help is necessary, then the number of Marines needed should be given after the command (e.g., *Support, two* or *Support, three*).
- Closing. Lets others know they are moving in depth or forward into the kill zone. It is usually given from a dominating position after the threat has been eliminated (e.g. when checking dead enemy). The other Marines within the enclosure adjust their sector of fire to support this action.
- Hold. The command, *Hold* is used to stop all actions within an enclosure and address a specific situation that could place fellow Marines in danger (e.g., a possible threat, danger area, dead space, or check to ensure whether suspected enemy dead are actually dead or not).
- Shot. This command is announced forcefully when a shot must be taken down the length of a hallway. This alerts others NOT to enter the hallway until they hear the command, *Continue.*
- Continue. Given only after the commands, *Shot* and *Hold.* This tells Marines they may continue with their actions.
- Danger area. This command identifies areas of concern for security purposes.
- Turn and go. This command relieves a Marine from a security position.
- Last room. This command informs the assault element that the limit of advance (i.e., the last room and location in a section of the objective) has been reached (e.g., *Last room, first deck*).
- Landslide. Directs all Marines to make rapid, controlled movement to a consolidation point.
- Avalanche. Directs all Marines to immediately evacuate the target site by whatever means available and move to a designated rally point outside of it. This is used only in emergencies, such as if a Marine discovers the structure is rigged to explode.

Additionally, units should establish SOPs for communicating linkup within a structure, as well as other potential events such as the discovery of IEDs or high value individuals/targets.

Clearing Procedures

The following sections represent best practices for clearing in the urban environment.

Rules of Room Clearing. Clearing buildings of enemy threats is an extremely dangerous task. Leaders should begin every urban clearing task by asking whether or not it is inherently necessary.

If a structure is believed to be fitted with booby traps and it holds no real tactical value, then it may be bypassed as long as it is covered by a security element. If a structure is undeniably hostile and there is a low chance of civilian casualties, then the best way of clearing it would be through the use of fire support assets. Marines should only be committed to close quarters clearing when the mission requires it. When it does, leaders should ensure Marines have sufficient fragmentation and/or flash bang grenades to dominate confined spaces. Rules for clearing a room are as follows:

- Never clear with a Marine what could instead be destroyed with high explosives.
- Doors should be used as a last resort. (When possible, make your own entry point.)
- When able, prepare all rooms with fragmentation or flash-bang grenades prior to entry.
- Never clear a room alone.
- Avoid throwing grenades up a staircase.
- Not all walls afford equal protection; know what cover is available in the urban environment.
- Take accountability upon entering and exiting each room.
- Never stack in open exterior areas.
- Overwatch walled compounds prior to entry.
- As soon as there are no eyes on a room or structure, it cannot be assumed to be clear; back clearing may be necessary.
- Urban fighting is three-dimensional—clear right, left, upward, and downward.

The primary consideration in interior clearing is utilizing the geometry and cover provided by the structure to the advantage of the assault element. As a general rule, Marines should strive to engage targets "one room over" rather than blindly rushing into rooms (see figure 6-19). The standard clearing sequence is—

- Clear as much of the room as possible from the exterior.
- Clear the fatal funnel.
- Dominate the room.
- Eliminate threats.
- Control the room
- Hasty search.
- Mark.
- Evacuate.

Clear as Much of the Room as Possible from the Exterior. Marines should use the building's geometry to their advantage by employing the "pie method" to limit their exposure to danger areas while methodically clearing threats from the objective.

Clear the Fatal Funnel. The entryways to rooms, hallway junctures, and breach points are natural aim points for defending adversaries. When making entry, Marines must aggressively move past this "fatal funnel" of incoming fire.

Warning
NEVER stop in the fatal funnel for any reason.

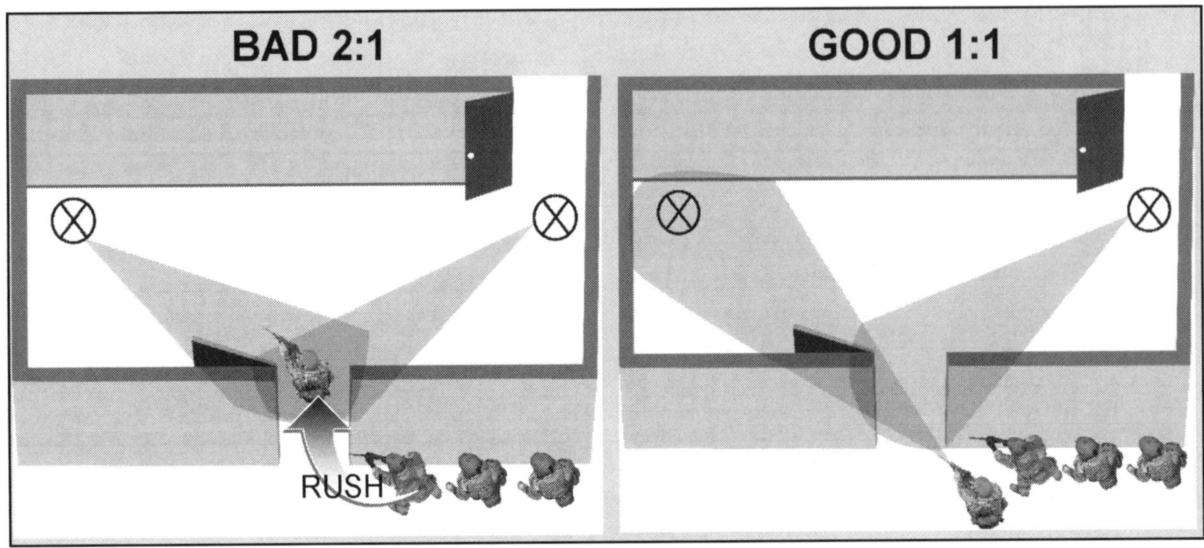

Figure 6-19. Clearing From One Room Over.

Dominate the Room. Upon entry, seek out and secure dominant positions that offer the best coverage of the room and ensure they have open and interlocking fields of fire.

Eliminate Threats. Systematically neutralize threats in the room through violence of action.

Control the Room. The senior Marine dictates follow-on actions once threats have been eliminated. No further movement through the room is authorized without the assault element's approval.

Hasty Search. Investigate the room and clear any secondary threats in the following order:

- Search the living.
- Search the dead.
- Search the room.

Mark. The room is marked to indicate it was cleared and searched according to the unit's SOPs. However, if a space is marked and no Marine has had constant observation on it, it must be re-cleared if the unit intends to traverse it again.

Evacuate. The assault element gains accountability and leaves to continue clearing the building.

Single Room Clear. When clearing a single room, the number one Marine in the stack identifies a room that has not been cleared and "pies" off the entryway a step at a time, thus clearing the room from the outside. Once as much observation of the room as possible has been gained from the outside, the Marine enters and goes to the least observed corner of the room. The number two Marine follows through the doorway close behind and moves toward the opposite direction, then orients on the center of the room or the next area that cannot be observed (see figure 6-20).

The remaining members of the stack enter the room and pick up sectors of fire toward the center of the room (see figure 6-21). The team then either prepares to continue clearing forward or exits the room, with the team leader getting accountability as the team exits.

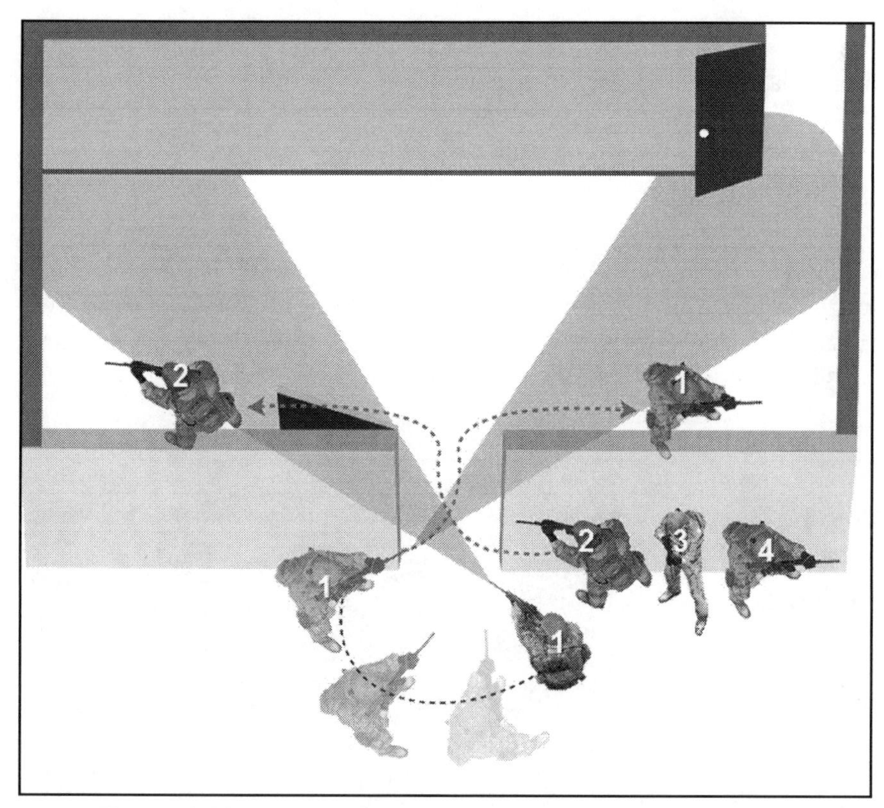

Figure 6-20. Number One and Two Marines Make Entry.

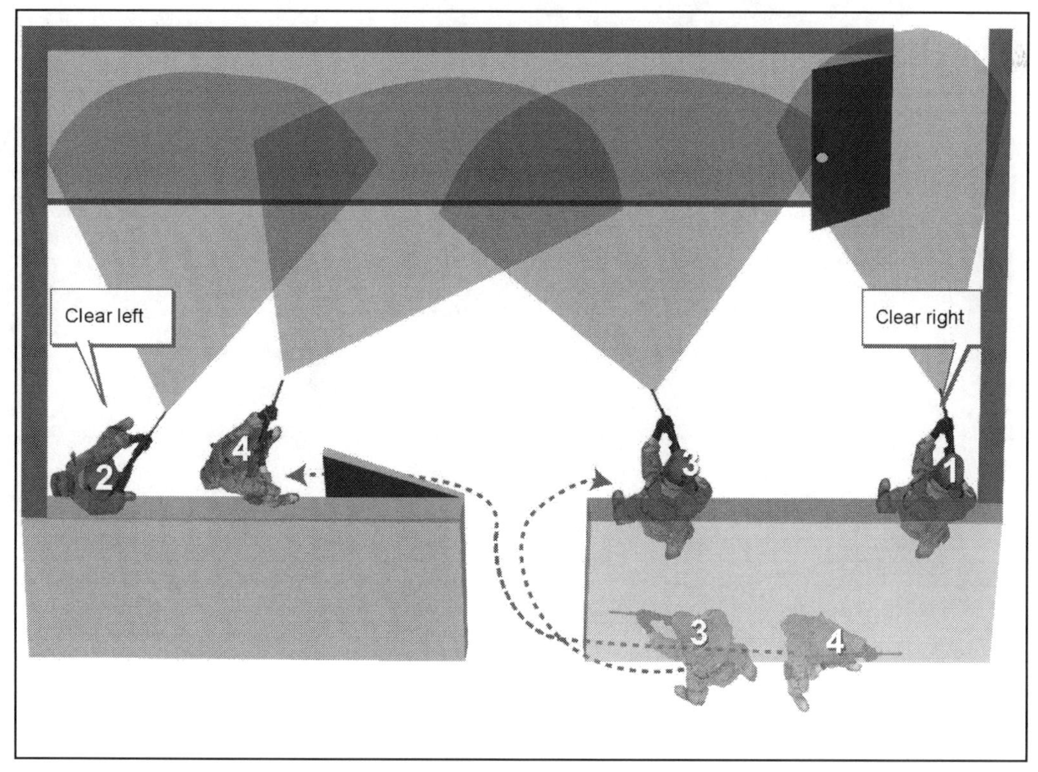

Figure 6-21. Number Three and Four Marines Enter the Room.

Clearing with Support. After a room is entered, it may be necessary to call for support due to casualties, an interior layout which requires more security, or other unforeseen reasons. In this case, the senior Marine in the room uses the command, *Support*, followed by the number of Marines needed to fulfill the requirement. The Marines entering the room announce, *(number) Marines entering* in order to prevent fratricide (see figure 6-22).

L-Shaped Hallway Intersection. If an L-shaped hallway is encountered, Marines use similar principals to room clearing. The number one Marine calls out the hallway once they recognize it and stops the forward motion of the stack just short of the hallway turn, taking care not to let the body or rifle barrel protrude past the hallway corner (see figure 6-23).

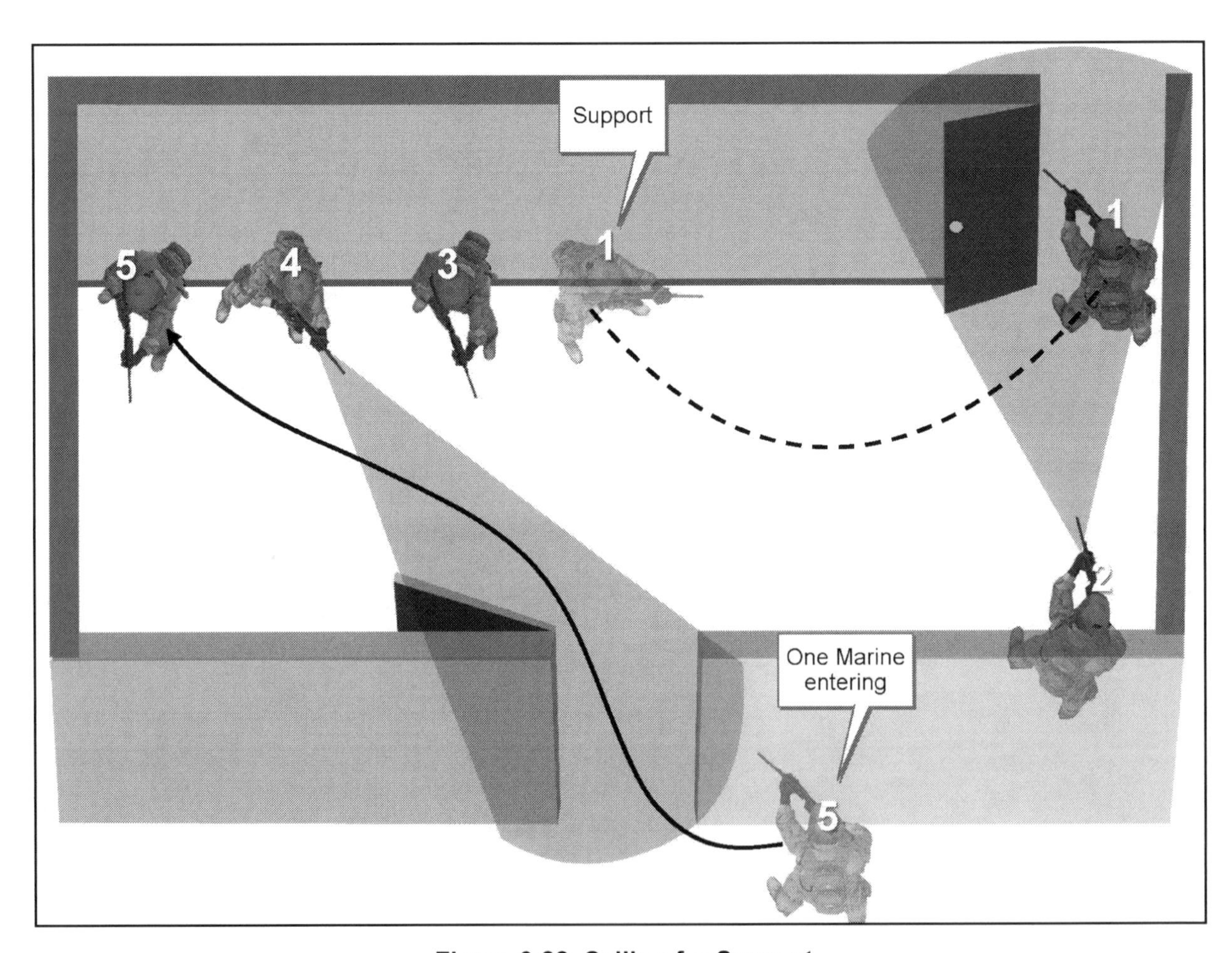

Figure 6-22. Calling for Support.

Figure 6-23. Number One Marine Identifies Intersection.

The number one Marine holds forward security while the number two Marine begins to pie off the hallway corner (see figure 6-24).

Once the number two Marine has "pied" off as much as possible without protruding past the hallway corner, they signal the number one Marine and they will both "pop" around the corner simultaneously, with the number one Marine holding close to the corner and using it as cover while the number two Marine moves aggressively to get clear of the number one Marine and dominate the long axis of the hallway (see figure 6-25).

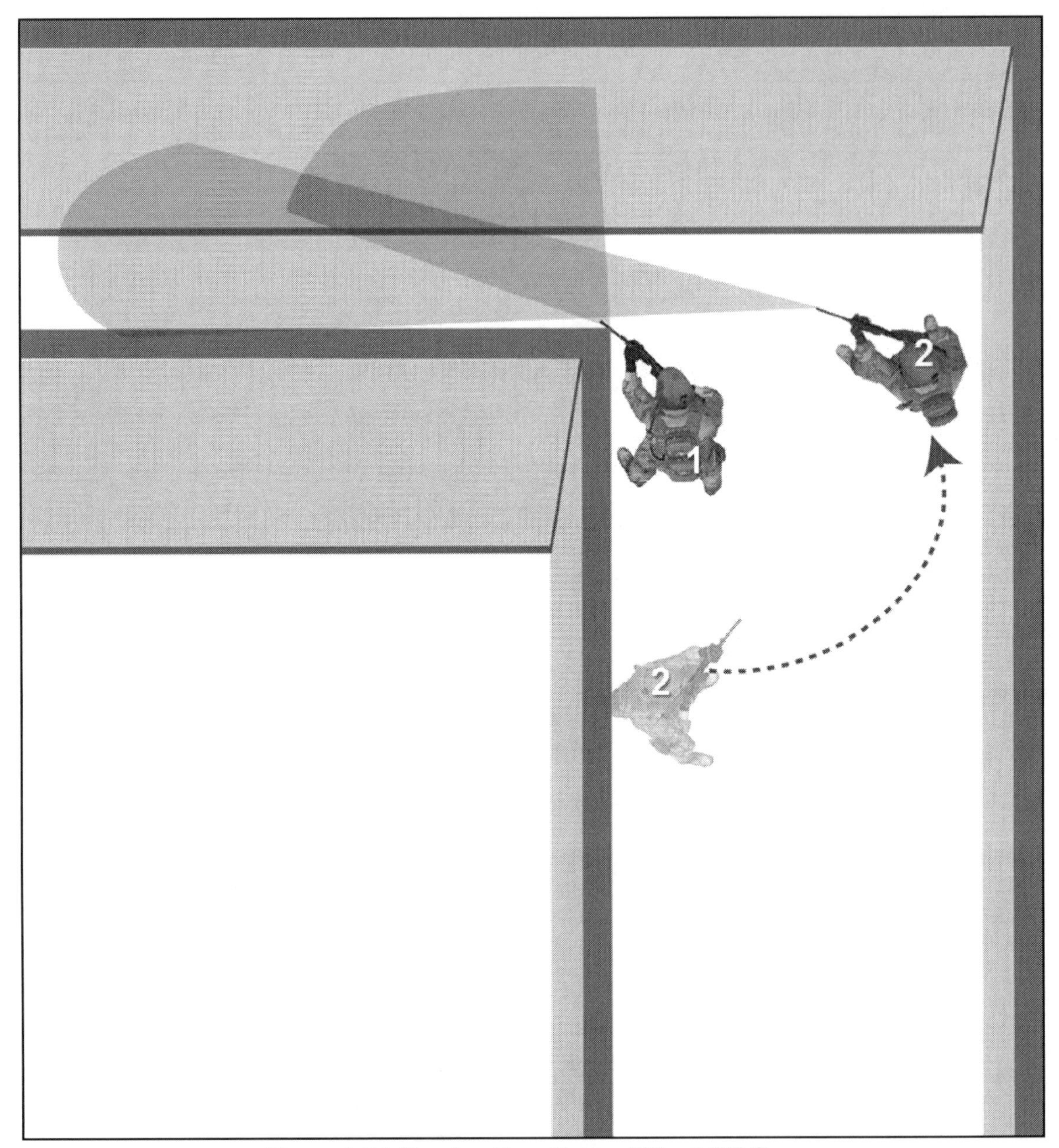

Figure 6-24. Number Two Marine "Pies" Off Corner.

T-Shaped Hallway Intersection. The procedures for T-shaped hallways are similar to L-shaped hallways. In this case, the number one Marine calls out the hallway and moves to one of the corners, picking up security on the hallway opposite. The number two Marine follows suit and takes up a position on the opposite wall (see figure 6-26).

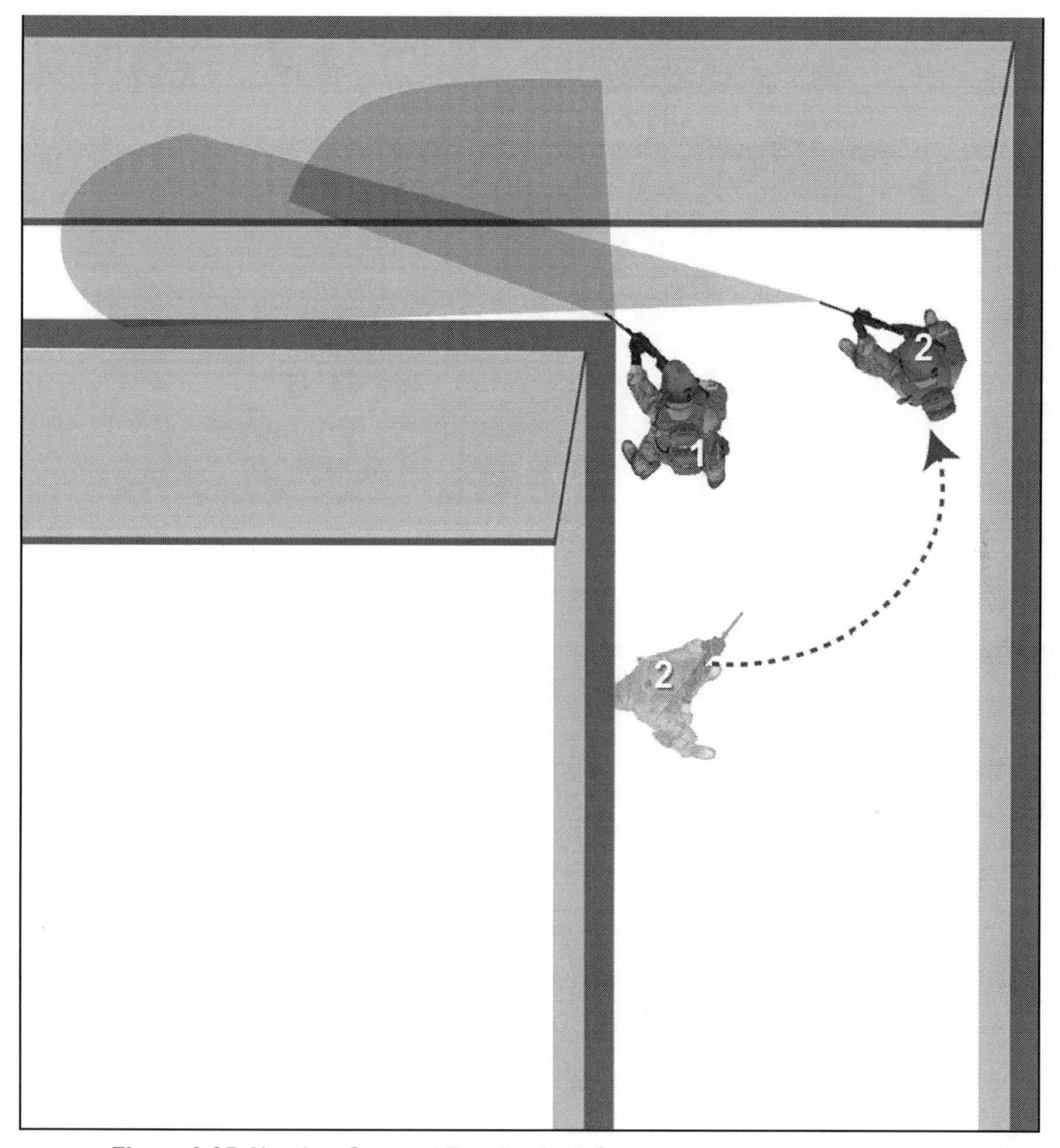

Figure 6-25. Number One and Two Marines Simultaneously "Pop" the Corner.

On signal, both Marines simultaneously pop their corners, switching sectors of fire and dominating the long axis of the hallway, utilizing the corner as cover if possible (see figure 6-27).

Stairwells. After entering a building, the assault element may encounter stairwells. Stairwells are the most difficult areas of a building in which to operate. Knowing the building construction would give the assault element a significant advantage in determining which technique to use when clearing a stairwell. There usually two types of stairwells—continuing and noncontinuing.

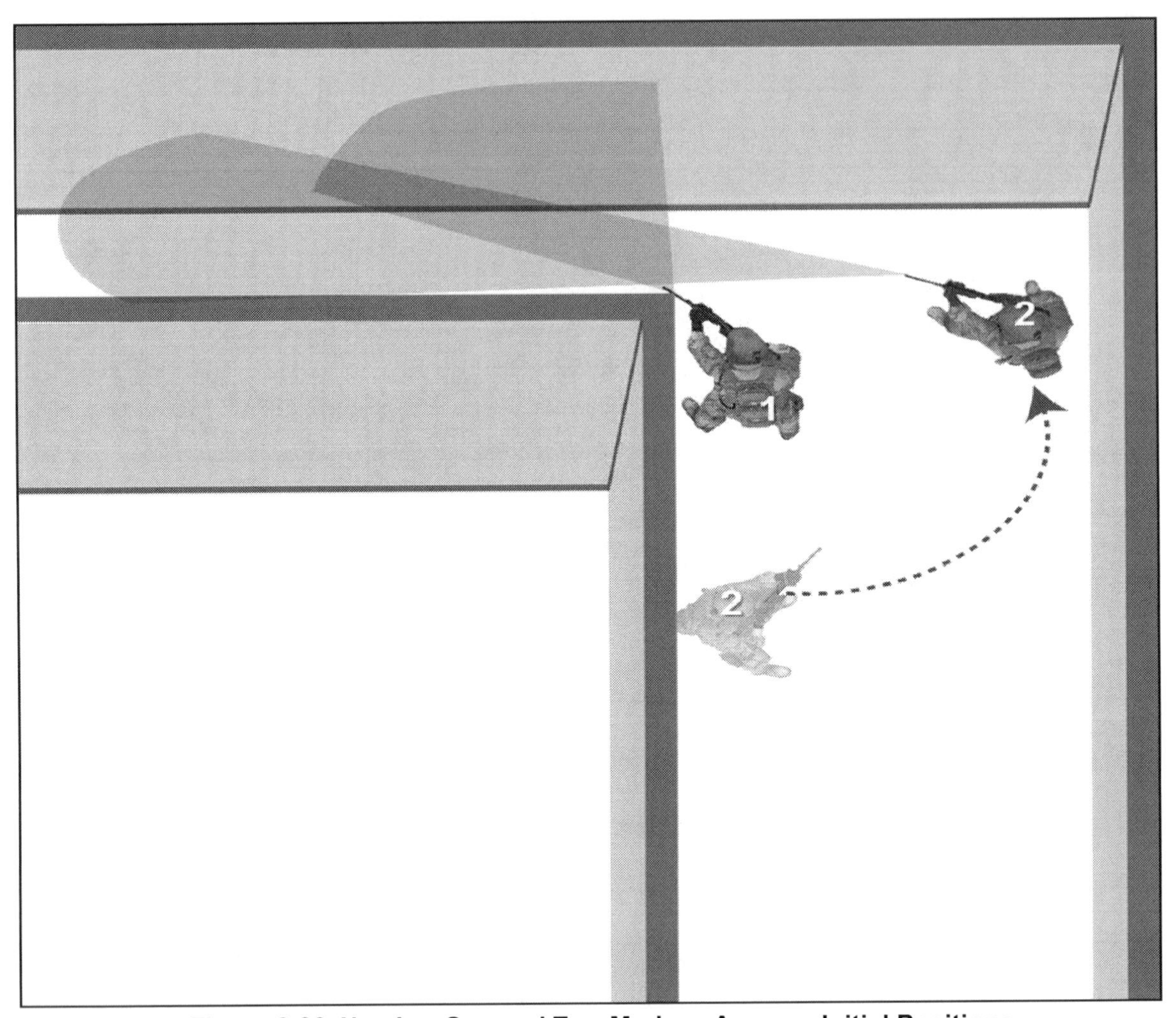

Figure 6-26. Number One and Two Marines Assume Initial Positions.

Continuing stairwells (i.e., switchbacks) are typically on the ends or at the corners of buildings, and usually close to elevators. Continuing stairwells should be cleared in segments by "bounding" rapidly between landings while maintaining security under landings, overhead, and to the front and rear. Teams move progressively up or down the stairway levels. Once a flight of stairs has been cleared, the clearing team stops to maintain security on the door and to the front. The next team then assumes the assault team role and clears the next flight of stairs. This is repeated until the top or bottom floor is reached.

Noncontinuing stairwells are more open and have two or fewer landings. They are usually found near the center of buildings close to large lobbies. Noncontinuing stairwells should be cleared through continuous movement by the original clearing team. Hesitating on the stairs makes Marines easy targets. Because these stairwells are open, security is very difficult to establish and maintain. Once security is established, one team moves to the next landing and sets security on the hallway. The next team then becomes the assault team to gain a foothold on that floor. Fatigue can be a major factor.

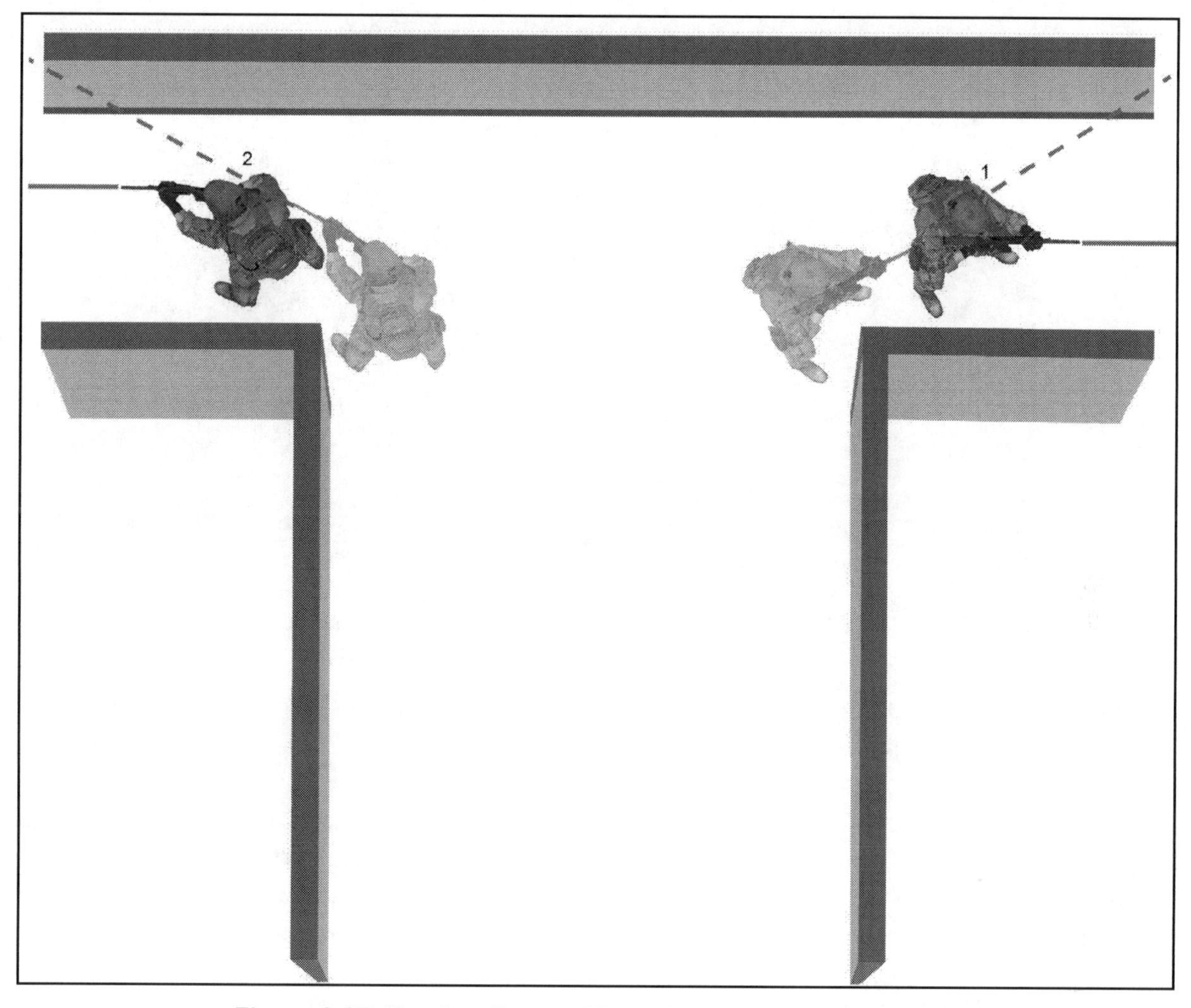

Figure 6-27. Number One and Two Marines Dominate a Hallway.

The following procedures are used for clearing noncontinuing stairwells (from bottom to top):

- The number one Marine leads up or down the stairs, one step ahead of the number two Marine. Upon reaching the point just before it is possible to be engaged from above or below, number one Marine turns around and covers overhead. From this point, the number one Marine ascends the stairs, moving backward while covering behind and above (see figure 6-28).
- The number two Marine follows the number one Marine upstairs, one step behind and to the side. When the number one Marine turns to cover overhead, the number two Marine remains oriented to the front, covering directly up the stairwell.

The clearing team's speed of movement is determined by the number one Marine. Marines pie as much of an area as possible before ascending each step.

Back Clearing. Due to limitations in force size, it is often necessary for a squad to conduct back clearing through previously cleared spaces. Since a space is only deemed clear as long as Marines have had continuous observation on it, any gap in observation requires a space to be cleared again.

Figure 6-28. Clearing Noncontinuing Stairwells.

Squad leaders should enforce this interior movement discipline in order to ensure Marines do not become casualties because of becoming too comfortable in their environment (i.e., complacency).

Upon entering a building, the squad leader should establish specific tactical control measures. A space or room should be designated as an immediate casualty collection point. Separate spaces should be designated for detainee searching (i.e., "dirty" site) and detainee collection and processing (i.e., "clean" site used after detainees are searched). Detainees should be kept separate from US casualties at all times. Additionally, should the situation require it, a space should be designated to collect all relevant site exploitation materials for further exploitation.

Marking. During assaulting and clearing, North Atlantic Treaty Organization (NATO) control markings must be used. These are characterized by the following:

- *Entry point*. A flag or luminescent light is used to indicate that the entry point is cleared of enemy forces and booby traps. It must be easily recognizable for approaching Marines (or other joint or coalition forces) during day or night.
- *Forward line of troops*. A flag or luminescent light is used to mark the clearing progress by building, floor, and room. Each room is marked as soon as it is cleared so supporting fires can be shifted for effects on the enemy and to reduce fratricide.
- *Building status/information*. The following colors are used to denote the building status:
  - Red–Building not yet clear.
  - Green–Building cleared.
  - Yellow–Casualties present.
  - Blue–Engineer support required.
  - White–Forward line of troops.

Defensive Considerations

The underlying principles and considerations for standard defensive operations in the urban environment are the same as in the offense. The primary difference lies in constructing a defensive position and in using urban terrain to the advantage of the defending force.

Considerations for urban defensive positions include—

- Firing positions should only be open enough to cover assigned fields of fire.
- Equipment and weapon muzzles should not protrude outside of buildings.
- Natural interior shadows should be used as camouflage.
- Structures should be ballistically and structurally reinforced (i.e., with sandbags, lumber, or other materials).
- Multiple exit points should be available for movement to alternate and supplementary positions.
- Items such as glass removed from occupied area to reduce secondary fragmentation

Considerations for urban rocket and missile employment include—

- Windows or walls must be removed from the back-blast areas of rocket firing positions.
- Some antiarmor weapons, such as those employing top-down attack vectors, may need a hot position in a space with no overhead cover. This hot position should be prepared with well-camouflaged frontal defenses and its employment should be rehearsed.

Considerations for designated marksman or sniper positions in urban environments include:

- Only the bare minimum opening in a structure should be cleared to allow for the traversal of the weapon and its optics over its assigned sector of fire.
- Suppressors should be employed when possible to reduce muzzle flash.
- The firing point should not be the only opening in the building; diversionary firing points should be used to conceal the Marine's true location.
- Designated marksmen and snipers should prepare and rehearse moving to and utilizing alternate positions in case the position is discovered.

Stability Actions

Cities are fundamentally designed to house large numbers of civilians. Because of this basic fact, Marines must mentally and tactically prepare to operate in an environment that is saturated with people, both innocent and threat. For further information on stability actions beyond what is included here, refer to MCWP 3-03, *Stability Operations.*

Tactical Callouts

Entering a structure for close quarters urban fighting is one of the most dangerous tasks a rifle squad may perform. It was described previously that the best way to clear a hostile structure is to destroy it without entering and risking Marines' lives. However, when there may be either combatants or noncombatants (or both) within a building, conducting a tactical callout to direct the orderly evacuation of noncombatants before clearing a structure is more appropriate and fundamentally safer than a straightforward urban attack.

Tactical callouts are characterized by deliberate planning, tactical patience, and vigilant security. During tactical callouts, Marines isolate the objective, conduct callouts, conduct detainee handling procedures, and deliberately clear the objective.

Isolate. Marines isolate the objective by employing an inner and outer cordon, much like in a cordon and search. Marines should ensure they have enough standoff from a structure to mitigate the risks of booby traps and rapid escalation into a close combat engagement.

Callout. Once the objective has been isolated, Marines will begin callout procedures (see figure 6-29) by gaining the attention of personnel in the objective. This can be done with flash bangs, a loudspeaker, pyrotechnics, or voice, as the situation dictates. Once Marines have the attention of personnel in the structure, commands must come from only one designated individual to the personnel within the objective for the duration of the callout.

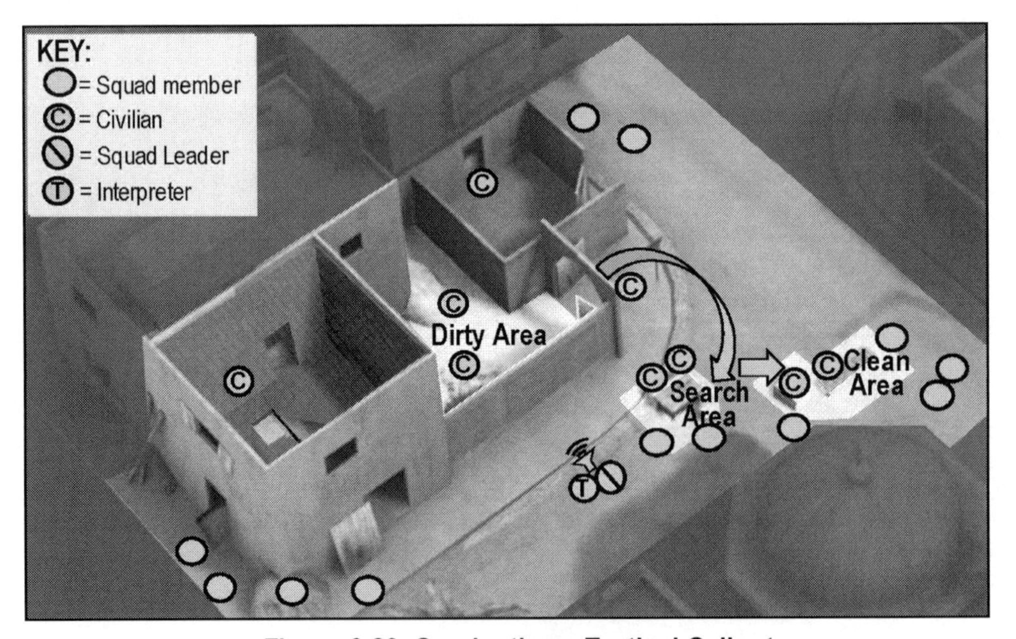

Figure 6-29. Conducting a Tactical Callout.

Marines explain the situation, establish the consequences of non-compliance, and issue clear, concise instructions to personnel within the objective. An example would be:

> *Attention! Attention! Attention! US forces have surrounded the building. Failure to comply with all instructions will put your life, the lives of your family, and your property at risk. If you attempt to use a weapon, you will be killed. Any attempts to escape will be treated as an act of aggression and you may be harmed. At this time, slowly move outside the building towards the sound of my voice with hands raised.*
>
> ***Note:*** All instruction should be given in both English and the local language (if other than English) to establish the Marine force's identity, and that they are not a criminal element or other nefarious actors.

Detainee Handling. The location where the building's occupants are to gather should be clearly marked. They should be directed to come forward to be searched one at a time. As they advance, optics such as thermal imaging should be used to scan for unseen threats, such as explosive belts under garments. The search team establishes a "dirty" area where unsearched personnel are searched and a separate "clean" area where searched personnel are gathered.

Once all personnel are searched, the unit should confirm with the detainees that the structure is now empty. Prior to proceeding to clear the structure, it may be helpful to temporarily send one detainee back into the building to open all doors and windows. This detainee should be instructed to do so rapidly and must be searched again upon exiting.

Clear. After confirming with the detainees that the building is clear, Marines deliberately clear it using the techniques discussed previously. The initial clearing of a building is to identify and neutralize any potential threats. Once that has been completed, Marines execute a deliberate site exploitation plan.

Once the objective has been cleared, all detainees have been cataloged, and site exploitation is complete, the Marines collapse their security from the objective, starting with the assault element and followed by the cordon. If other buildings must be searched, it may be useful to bring along civilian personnel from the first target building to facilitate communication and cooperation with personnel in follow-on buildings.

A tactical callout significantly reduces the risks to both Marines and noncombatants, and can be a very effective method of dealing with a population that Marines do not wish to alienate through excessive violence. Employing them requires a high degree of tactical patience. As with most other elements of urban warfare, slow is fast. Hurried tactical actions in the urban environment greatly increase the risks to Marines.

Checkpoints

Checkpoints can be established either to ensure security from vehicles or personnel entering and exiting an area, or to deny access and movement. During stability actions, checkpoints are used to control the movement of the general public and deny the enemy freedom of movement. During traditional offensive and defensive operations, checkpoints may also be used to account for and control the movement of friendly forces and equipment. Checkpoints are used as tools to check vehicular and pedestrian traffic to enforce movement control measures and other laws, orders, or regulations. When applied to combat operations, checkpoints typically involve a detailed search of people and vehicles in a safe area covered by observation and fires. They can be hardened for more permanent checkpoints or hastily established for temporary operations. An established checkpoint is an efficient way to display a show of force in areas habitually targeted by enemy forces. All Marines must understand how to establish, plan, and execute effective checkpoint operations in order to contribute to the security of friendly units, as well as noncombatants.

Checkpoint Planning. Any ground unit may be called upon to establish permanent checkpoints at road junctions, bridges, trails, borders, outskirts of cities, or on the edge of controlled areas. These can be set up to optimize force protection and still be effective. On the other hand, a hasty or temporary checkpoint can be used for spot checks. It can consist of two or more vehicles placed diagonally across a road, a coil of barbed wire, or just traffic cones. Its strength is the element of surprise. Thus, it is most effective within the first half hour of being in position. When possible, it should be placed where it cannot be seen from more than a short distance away, such as sharp bends, dips in the road, or over a hill or prominent terrain feature. Once word of its presence spreads through the local population, its tactical value goes down, while its force protection requirements go up; there is only a small likelihood that a sophisticated threat will stumble into one of them. Commander's guidance, unit SOPs, and theater rules of engagement provide additional guidance that may be required to plan checkpoints, such as the use of female searchers, interpreters, the force continuum, and appropriate signs to convey instructions in the native language and/or dialects.

Checkpoint Missions. Checkpoints can be established for a variety of short and long-term mission requirements. These missions typically include the following:

- Control vehicles and people in order to prevent large crowds from assembling.
- Assist in locating persons of interest for detention, capture, or questioning. Biometric identification devices can also be employed to register inhabitants and/or confirm their identities.
- Assist partnered forces in enforcing their laws and assist in legitimizing their government.
- Dominate the area around the checkpoint. A checkpoint with local patrolling around it shows force and assists in maintaining law and order.
- Prevent the movement of contraband by threat forces, such as weapons and explosives.

Vehicle Checkpoint. A vehicle checkpoint (VCP) is established at a designated location on the ground, a waterway, or a road or trail network used to control and influence the flow of pedestrian, vehicular, or boat traffic to execute tactical tasks and generate effects. A VCP can be hasty or deliberate in nature. Its purpose may be friendly, terrain, enemy, or environmentally oriented.

Location. The locations of VCPs are often based on natural or logical divisions between different ethnic, political, and/or religious factions. However, when possible, units should try to—

- Establish VCPs at crossroads to deny the enemy freedom of movement in several directions.
- Establish VCPs at least 100 meters away from sensitive buildings to deny intelligence on the building to any potential threat, and reduce effects on building of bomb detonated at checkpoint.
- Locate entry and exit checkpoints side by side, unless they are placed on one-way streets.

Set-Up. Speed bumps and speed limit signs generally have little tactical value, but are useful for determining a driver's intent. Figure 6-30 gives an example of a VCP. The following are some general guidelines for setting up a VCP:

- A security team or reaction force of appropriate size for the threat should be positioned to support the VCP.
- The vehicle being checked should be isolated from other cars by a barrier of some type or a Marine that regulates the flow of traffic.
- Signs should be used to advise drivers of the checkpoint approximately 100 meters ahead of the VCP. Signs should be in both English and the language of the local populace, if possible. Additionally, having two signs—one in written language and one using easily understood

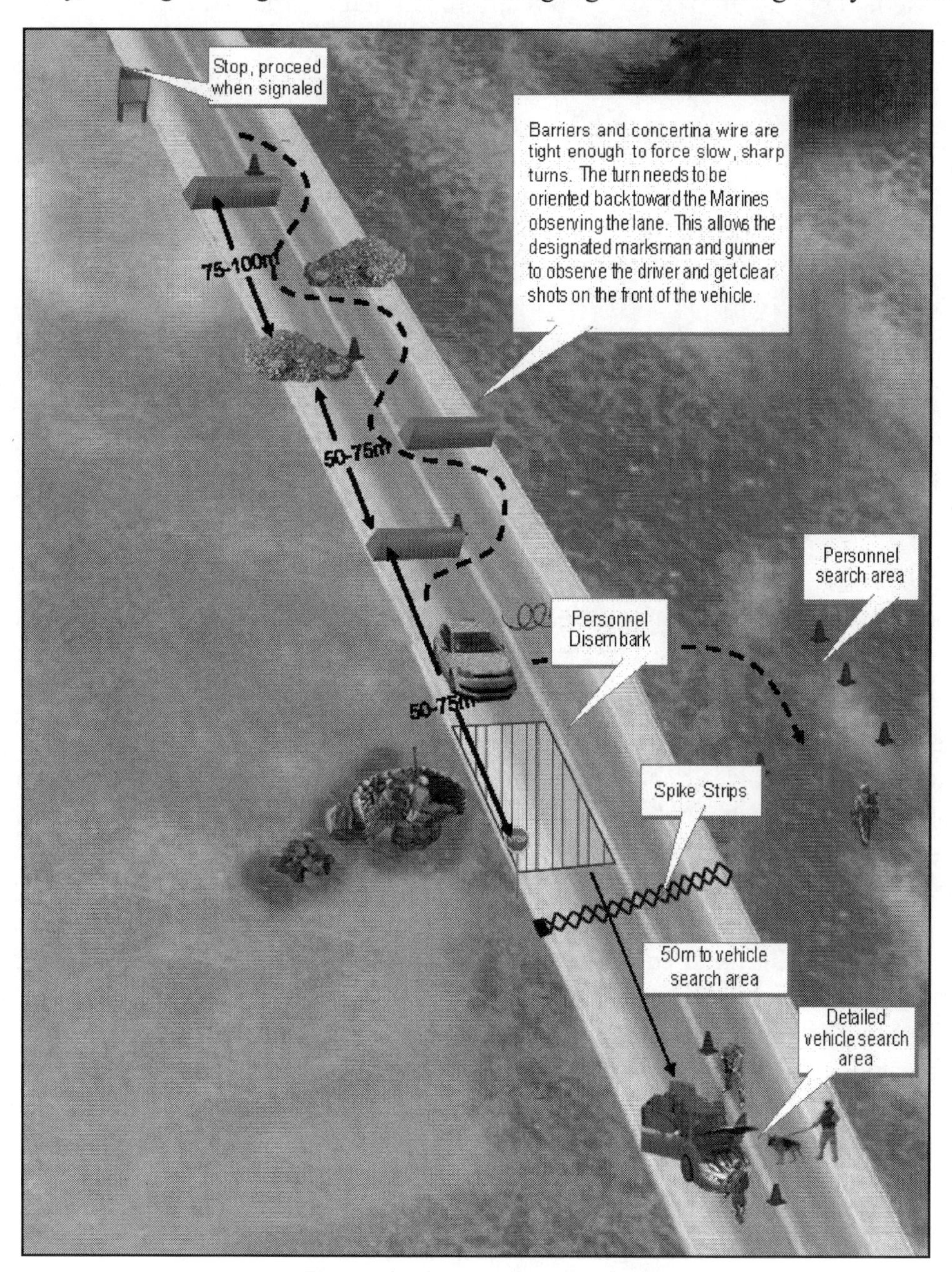

Figure 6-30. Vehicle Checkpoint.

traffic symbols—doubles the effort of protecting Marines and the local populace from misunderstanding and intent when coming up to a VCP.

- Vehicles should be canalized without an alternate way in or out until they have the consent of the Marines running the checkpoint to continue.
- A suitable crew-served weapon should be positioned on the opposite side of the Marine making the check in an elevated position covering the driver. The crew-served weapon gunner should be able to see exactly what the driver is doing at all times. A SOP or simple signal should be developed that allows the immediate employment of this weapon when necessary.
- All other Marines should be on the same side as the crew-served weapon to avoid possible fratricide. However, not all security personnel should be obvious, and overwatch and reaction positions should be shifted frequently to retain functional surprise and provide effective maneuver options.
- Employ overwatch whenever possible and be ready to prevent or counter any attack.

When dealing with VCPs, squad leaders should ensure their Marines are well versed in the current rules of engagement and force continuum procedures. The force continuum is defined as sequential actions which begin with nonlethal force measures and graduate to lethal measures, to include warning, disabling, or deadly shots in order to defeat a threat and protect the force. (For more information, refer to MCTP 10-10A, *Multi-Service Tactics, Techniques, and Procedures for the Tactical Employment of Nonlethal Weapons [NLW].*) The following is an example of gradually applying force to a situation:

- An audible warning should be used to alert people to the presence of Marines (e.g. horn, air horn, loudspeaker, flash/bang device, or siren).
- Visual aids should be used (e.g., lights, chem lights, flares, colored flags, or signs).
- The appropriate weapons posture and intent to use it should be demonstrated.
- Nonlethal means should be attempted (e.g., stop strips, physical barrier, vehicle, visual/audio signal, or signs).
- Warning shots should be fired.
- Disabling shots should be fired (e.g., at the tires, engine block, and windows) to attempt to stop the vehicle. It is a good idea to have a Marine designated for this job so that not all Marines end up shooting the vehicle and possibly causing harm to a civilian not paying attention.

Use of Deadly Force. The rules of engagement and force continuum procedures can change daily and from area to area. Squad leaders must ensure Marines are briefed and that they review them before every mission. After a force continuum incident and when back at a secure location, Marines should be debriefed on what happened and what went well or poorly. Keep in mind that the force continuum does not pertain only to checkpoints; it applies to convoys, patrols, cordons, and other missions. The procedures are important to know and understand in order to save the lives of both friendly forces and civilians.

Verifying Procedures. The following procedures are used at a VCP to verify that a vehicle is safe and should be permitted to pass:

- To the maximum extent possible, the same Marines (i.e., specifically trained) should be used to approach the vehicle.
- The number of people in the car should be counted before approaching it.
- From a covered position, the driver should be signaled to open their window and shut off the vehicle engine.
- The driver should be directed to release the hood and trunk if it can be done from inside the car.
- The driver's identification or papers should be requested.
- If the driver or any occupant is identified as a suspect in some irregularity, a signal for the rest of the security team should be given that will not alert the occupants of the car.
- The driver should be directed to get out of the car and open the hood and trunk fully.
- Marines should stand where they can clearly observe the driver.
- The driver should be ordered to take a knee and raise their hands where they can be seen by the other Marines.
- This procedure should be repeated one at a time with each vehicle occupant until it is empty.
- All passengers should be escorted to a designated search area off the road. The driver is searched first and then allowed to return to the vehicle and move it to the detailed vehicle search area. All other passengers are not allowed to return to the vehicle until the more detailed vehicle search is complete.
- If the driver is detained, do not let their friends take the car they were driving.
- Everything found in the search should be turned over to the proper authorities.

When appropriate, local police should be present to make arrests. It is usually best to have all passengers (not the driver) taken to a holding area away from the car during the search. In the event vehicles refuse to stop or appear intent on "running" the checkpoint, caltrops, or spike strips that will puncture tires should be used. These can be easily and rapidly scattered on a road as a nonlethal device to deter attempts to run the VCP. If these are not available, Marines can use nails driven through lumber as hasty nonlethal vehicle stopping devices.

Checkpoint Documentation. Marines should make every effort to document people and vehicles passing through the checkpoint. This is particularly useful for creating a census of the people in the area of operations and tracking people's movements. This also assists in understanding how people move within and in and out of the area of operations, and for what reasons.

Personnel Checkpoint. Personnel checkpoints (PCPs) are used to regulate pedestrian traffic. They can be used on a systematic or random basis. While these do not involve checking vehicles, they are used for the same purpose as the VCP.

Systematic Personnel Checkpoint. These are permanent or semipermanent positions. They are used when routine, methodical checks of all persons are required. For example, a systematic PCP could be used to control access to a food distribution site. They are also used to ensure the security of potentially vulnerable areas, such as at access points to a headquarters element. The following procedures for this type of checkpoint have been found to be effective:

- All personnel should be informed what papers are needed to cross the line. Signs and media announcements should be used as appropriate so people know what to expect.
- Barriers should be used to regulate the flow of people rather than signs.
- If security overwatch can be maintained, more than one line should be used.
- The number of people immediately in front of the PCP should be minimized.
- Maintain at least 10 meters of standoff distance between the PCP entry and the head of the line waiting to enter. As with VCPs, Marines should establish easily understood signals between the squad leader and the overwatch team so they can respond quickly to any emergency.

Random Personnel Checkpoint. The random nature of the checkpoint can refer to the places that the checkpoints are established, in which case they are probably not permanent or semipermanent positions. It can also refer to the way that personnel are screened. In a randomly placed checkpoint, procedures can be just as thorough as for the systematic PCP described above. On the other hand, if the checkpoint is used to screen some—but not all—people passing through a point (e.g., for weapons, drugs, or other contraband), its characteristics are as follows:

- Marines must have guidance on how to select those to be checked.
- It should not interfere with a reasonable flow of people.
- It must be in such a location so that a person cannot escape the checkpoint if confronted.
- It can take fewer Marines to operate. An overwatch team is still necessary for emergency reaction.

Whenever possible, small streets/alleys should be used to canalize pedestrians.

CHAPTER 7
MOVEMENT

MOUNTED OPERATIONS

Marine infantry squads are foot-mobile by design. When required, task-organized infantry squads may conduct tactical activities across the range of military operations mounted in vehicles that are wheeled or tracked, armored or unarmored, or some combination thereof. Employing combat vehicles with infantry squads increases their combat power. Regardless of the vehicles used, the current family of Marine Corps vehicles in which an infantry squad may be mounted are not infantry fighting vehicles, and should not be employed as such. They lack sufficient armor protection, stabilized weapons stations, means for the infantry to fight from the vehicle without exposing themselves to direct fire, and possess low silhouettes.

Integrating combat vehicles with infantry squads combines the advantages of the vehicles' mobility, firepower, protection, and the information platforms available while also increasing the squads' ability to operate in restricted terrain.

Based on considerations such as mission requirements, logistical support, length of the mission, and the armaments and vehicle assets available, squads may receive support in three ways: internal, external, or cross-attachment (see table 7-1).

Table 7- 1. Mounted Support.

Internal	External	Cross Attachment
Squad physically possesses a suite of vehicles to support the entire squad. Squad usually provides its own trained drivers. Squad conducts its own maintenance and other logistical operations (fueling, preventative maintenance, etc.). Squad must be proficient on crew-served weapons.	Outside unit provides vehicle support for the squad (e.g. truck company, AAVs, etc.). Requires development of positive action and teamwork between the squad and supporting unit. Supporting unit provides personnel and maintenance requirements.	The most common variation of cross-attachment the infantry squad will experience is that of infantry and armor platoons or companies between infantry and tank battalions, resulting in a tank-heavy task force and an infantry-heavy landing team.

Combat Vehicles Supporting the Squad

Combat vehicles give support by leading the squad in open terrain and providing one or more fast-moving mobile weapons systems. They can suppress and destroy enemy weapons, bunkers, and armored vehicles by fire and movement. They may also provide transportation when the tactical situation permits. Combat vehicles support the squad with mobility, firepower, and protection.

Mobility. Combat vehicles provide mobility support to the squad as follows:

- Assist opposed entry into bunkers or buildings.
- Breach or reduce obstacles.
- Provide transportation to the squad.
- Perform CASEVAC and resupply functions.

Firepower. Combat vehicles provide firepower support to the squad as follows:

- Lethal and accurate direct fire support.
- Suppress identified enemy positions.
- Acting as a mobile base of fire to provide a high volume of suppressive fire.
- Provide accurate fires while the vehicle is moving with stabilized gun systems.
- Destroy enemy tanks and other armored vehicles.

Protection. Combat vehicles provide protection to the squad as follows:

- Protect squad movement over open terrain with limited cover and concealment.
- Establish roadblocks or checkpoints.
- Provide limited obscuration with mounted smoke grenades.
- Isolate objectives with direct fire to prevent enemy withdrawal, reinforcement, or counterattack.
- Secure cleared portions of an objective by covering avenues of approach.

Squad Support to Combat Vehicles

The squad provides support to the combat vehicles through dismounted activities, including locating and breaching or marking antitank obstacles. The squad detects and suppresses or destroys enemy personnel armed with antitank weapons. Squad and fire team leaders may designate targets for vehicles and protect them in urban or close terrain.

Mobility. Other functions that the squad may provide in mobility include—

- Seize and retain terrain.
- Clear defiles and restrictive terrain ahead of vehicles in close or urban terrain.

Firepower. Other functions that the squad may provide in firepower include—

- Clear trenches, bunkers, and enter and clear buildings on objectives ahead of vehicles.
- Employ antitank weapons systems to destroy armored threats.

Protection. Other functions that the squad may provide in protection include—

- Provide local security of dead space or blind spots that weapons systems' or vehicle crews' fields of view cannot cover.
- Consolidate; perform detainee handling procedures and direct CASEVAC.

Capabilities

The primary role of combat vehicles is to provide the infantry squad with mobility to allow them to maneuver. Combat vehicles may also provide bases of fire, protection, breaching capabilities, enhanced communication platforms, and a variety of support functions that include resupply and CASEVAC capabilities. Effective integration of these units provides complementary and reinforcing effects.

There are three general reasons for employing combat vehicles with infantry squads:

- So the combat power capabilities of the vehicle can support the squads' maneuver.
- So the infantry squad can support combat vehicle maneuver.
- To achieve mutual support.

Technical Capabilities. Infantry leaders must have a basic understanding of the technical capabilities of combat vehicles. These include the vehicle characteristics, firepower, and protection. Squad leaders must have a clear understanding of the capabilities and limitations of their equipment. The amphibious assault vehicle (AAV), light armored vehicle, and high mobility multipurpose wheeled vehicle (HMMWV) each have their own capabilities, limitations, characteristics, and logistical requirements. Even though their role to the infantry is virtually the same, these vehicles provide support in different ways. To effectively employ combat vehicles, squad leaders must understand the specific capabilities and limitations of the vehicles that may be attached to their units. The following information, including table 7-2, is a brief overview of the combat vehicles' characteristics and combat power.

Table 7- 2. Vehicle Characteristics.

Specification	HMMWV	AAV	LAV-25
Tracked/Wheeled	Wheeled	Tracked	Wheeled
Length	196.5 in.	321.32 in.	252 in.
Width	86 in.	130.61 in.	98 in.
Height	74 in. (without crew-served weapon and turret)	130.56 in.	106 in.
Weight	5,600 lbs	46,314 lbs (unloaded)	25,600 lbs
Max Speed	78 mph	45 mph Land/8 mph Water	62 mph Land/6 mph Water
Load capacity	5 (with crew-served (weapon gunner)	21 Marines/ 10,000 lbs cargo.	4 Marines

Armament. The weapons and ammunition of combat vehicles are designed to defeat specific enemy targets, though many are multi-purpose. A squad leader with a basic understanding of these weapons and ammunition types will be able to better employ combat vehicles to their maximum effectiveness to defeat the enemy. Table 7-3 lists the basic weapon and ammunition types generally available to support the infantry squad.

Table 7- 3. Weapons and Ammunition.

	HMMWV		AAV		LAV-25	
	Weapon/Ammo	**Target**	**Weapon/Ammo**	**Target**	**Weapon/Ammo**	**Target**
Blast Munition	**40-mm MK 19** Max area: 2,212m Max point: 1,500m	trucks troops bunkers buildings	**40-mm MK 19** Max area: 2,212m Max point: 1,500m	trucks troops bunkers buildings	None	
Cannon	None		None		25-mm Bushmaster cannon	Light armored vehicles fortifications
Machine Guns	**M240B 7.62-mm** (mounted) Max area: 1,100m Max point: 800m **M2 .50 caliber** Max area: 1,830m Max point: 1,200m	troops, trucks, equipment	**M2 .50 caliber** Max area: 1,830m Max point: 1,200m	troops, trucks, equipment	**2 X M240 7.62-mm MG** 1 x coaxial mounted 1 x Turret mounted	troops, light vehicles, equipment
TOW Missile	Max effective: 3,750m	tanks bunkers	None		**LAV-ATvariant,** twin mounted missile carrier, 3,750m range	Armored vehicles, bunkers, fortifications
Breaching	None		**Mk-154 Linear Mine Clearing Kit** breach:100m long 16m wide 95% proofed	minefields & explosive obstacles	None	

LEGEND
AT antitank
LAV light armored vehicle
TOW tube-launched, optically tracked, wire-command link guided missile

High-Mobility Multipurpose Wheeled Vehicle

When supported internally, the squad may be equipped with a variant of the HMMWV, usually the armored M1151 variant. The HMMWV is a light (i.e., other than the M1151), highly mobile, diesel-powered, four-wheel drive vehicle equipped with an automatic transmission. Using components and kits common to the M998 chassis, the HMMWV can be configured to carry troops; armament; tube-launched, optically tracked, wire-command link guided missile launchers, or as a reconnaissance vehicle.

The HMMWV offers Marines many advantages and disadvantages in mobility, firepower, protection, and information.

The HMMWV provides the following mobility advantages to the rifle squad:

- The four-wheel drive chassis enables it to operate in a variety of terrain and climates.
- It is capable of fording 30 inches of water (60 inches if equipped with a fording kit).
- It can navigate narrow streets with minimal collateral damage.
- Some models employ a winch for the recovery of other similar-sized vehicles.

The HMMWV also has the following disadvantages in mobility:

- Its run-flat tires are susceptible to enemy fire.
- It possesses less ability to penetrate and breach defensive obstacles than tracked vehicles.
- It can be blocked by hasty or complex barricades.
- It may have a tendency to roll over on steep terrain (especially the M114 variant).

The HMMWV provides the following advantages in firepower to the rifle squad:

- It can employ a variety of crew-served weapon systems that provide direct fire support.
- Weapons systems can mount on all models with turrets, providing mobility and flexibility.

The HMMWV also has a disadvantage in firepower in that it can usually only mount one weapon system, making it less effective than other combat vehicles with multiple systems.

The HMMWV provides the following protection advantages to the rifle squad:

- The M1151 provides ballistic, artillery, and mine blast protection to occupants.
- The M1151 provides protection from 7.62-mm armor piercing rounds and air burst artillery.
- It provides limited antitank mine protection (12 pounds in front and 4 pounds to the rear).
- It provides 360-degree protection for the vehicle's gunner with its turret and shield.
- Armor packages are available that are effective against IEDs.

The HMMWV also has the following disadvantages in protection:

- All models other than the M1151 provide limited protection from direct and indirect fire.
- Gunners may be susceptible to direct and indirect fire in nonarmored variants.
- The lack of internal space present difficulties in conducting CASEVAC.

Some HMMWV variants have a variety of features that make them excellent for gathering and managing information. Some of these advantages include—

- The large front and side windows allow for excellent fields of view for vehicle occupants.
- They can carry a variety of communications systems.
- Amplifiers may be employed to extend the range of the communications systems.
- They can be configured to carry a variety of digital command and control devices.

- Crew-served weapons systems can employ optics with night vision, thermal, and range finding capabilities, including some with high resolution and magnification capabilities.

Some disadvantages it has in gathering and managing information include—

- Many of the digital, electronic, and communications systems require constant power sources. Therefore, the need to start the HMMWV to keep the batteries charged can present a tactical problem if stealth is desired during operations.
- The M1151 variant possesses a poor field of view due to its small front and side windows from the armor package.

Organization

Vehicle-mounted squads should be organized to maximize the capabilities and flexibility of their equipment and personnel. As with patrol organization, emphasis should be placed on maintaining fire team integrity. Although Marines should be cross-trained in the different vehicle duty positions, squad members must become highly proficient in their primary vehicle duty position in order to operate effectively and safely as a team.

Equipment. The mounted squad may be organized around three or four vehicles. While four is usually optimal for increased flexibility during operations, METT-T analysis drives the number of vehicles used. Typical modifications and enhancements include add-on armor kits, multipurpose power distribution systems, communication system mounts, GPS [global position system] devices, increased ammunition storage compartments, onboard power inverters, skid plates, jack plates, shoulder-launched munitions and light antitank missile mounts, rear equipment shelves with stowage boxes, and spare tire mounting systems.

The squad must be adequately equipped in order to accomplish its assigned missions. This often requires the ability to modify vehicles to best suit their missions and SOPs. The well-equipped mounted squad should have access to the following equipment at a minimum:

- Four tactical vehicles (M1151 variant minimum).
- Two trailers.
- Two M2 .50 caliber heavy machine guns.
- Two MK-19 automatic grenade launchers.
- Four M240B medium machine guns (if required).
- Thermal devices.
- Four vehicle-mounted radios.
- CREW [Counter radio-controlled improvised explosive device electronic warfare] systems.

Figure 7-1 and figure 7-2 display examples of a possible interior and exterior vehicle packing and load plan (minus individual equipment).

Personnel. The squad typically consists of 13 Marines and 4 vehicles. Although the squad will typically move and operate as a single element, it may on occasion operate as two sections (each consisting of six to seven personnel and two vehicles) or, for brief periods, the squad may

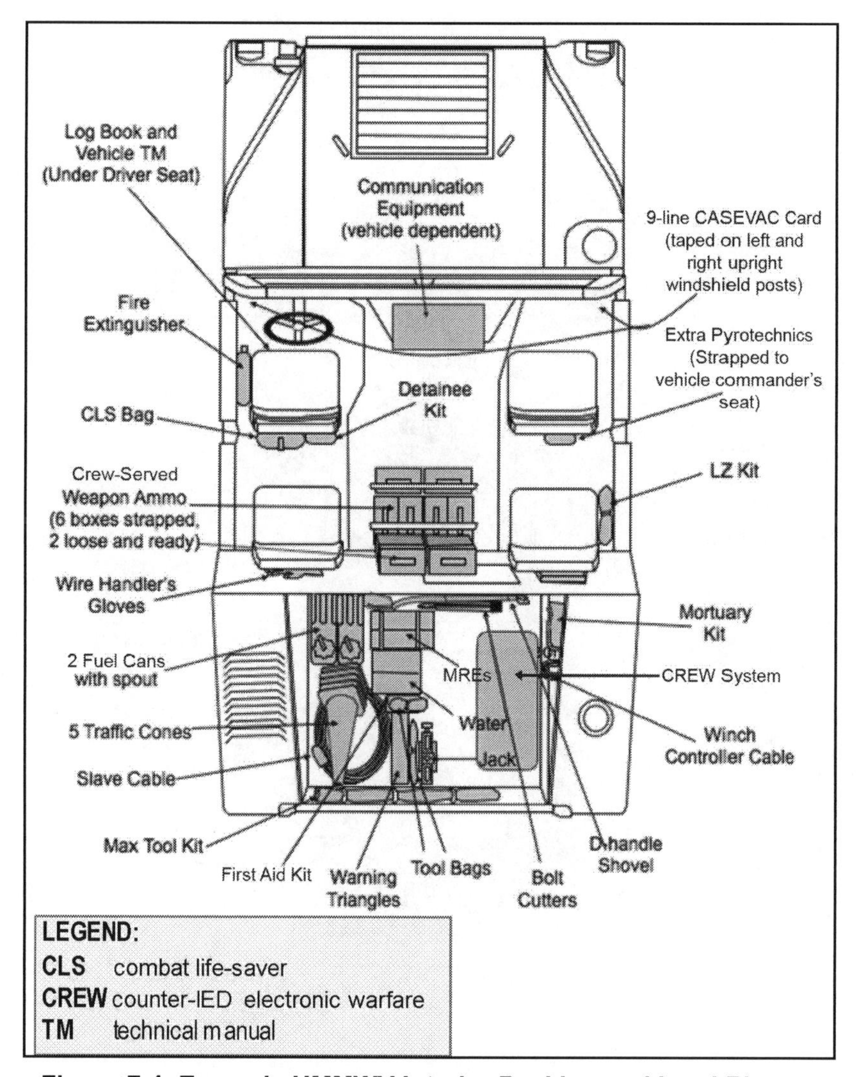

Figure 7-1. Example HMMWV Interior Packing and Load Plan.

further split into four independent elements (each consisting of three to four personnel and a single vehicle). When operating in this manner, care should be taken to ensure that each element can provide mutual support to the others. A minimum of three personnel should operate each vehicle (i.e., a driver, navigator, and crew-served weapon gunner). In order to account for contingencies, each Marine must be well versed in the duties and responsibilities of all duty positions. Squad leaders must consider how the family of mine-resistant, ambush-protected (i.e., MRAP) vehicles can be incorporated into operations and integrated with the other detachments when required.

> *Note:* Regardless of vehicle type; vehicles require at least three personnel for manning: a driver, navigator (i.e., vehicle commander), and crew-served weapon gunner.

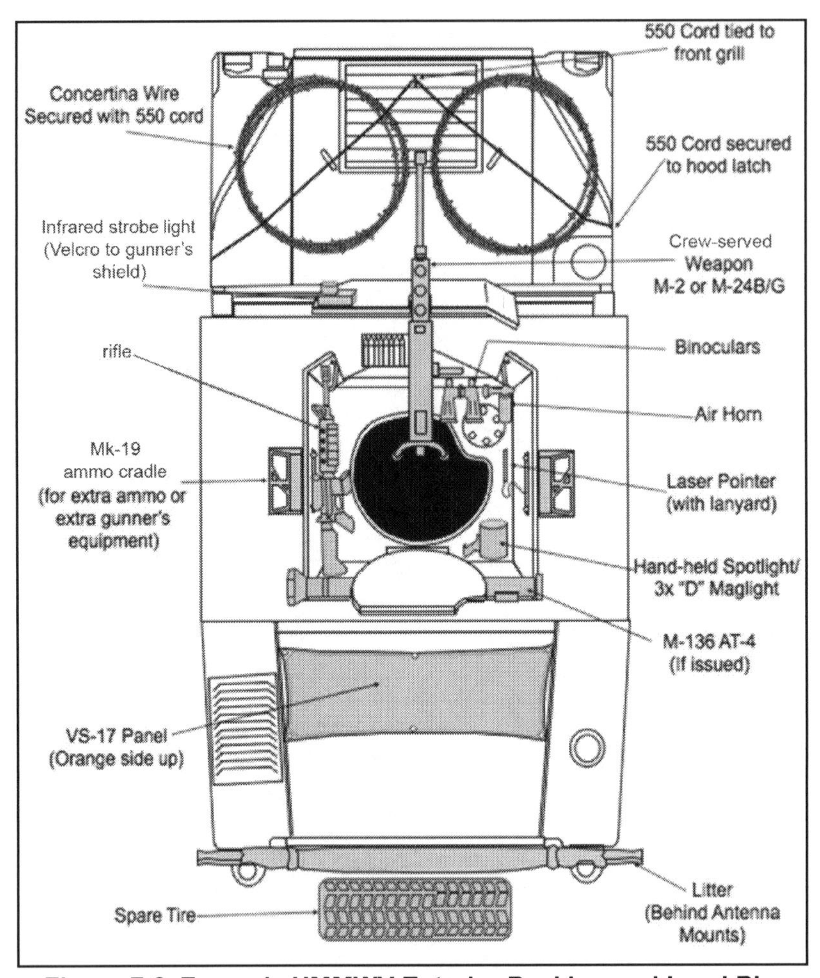

Figure 7-2. Example HMMWV Exterior Packing and Load Plan.

Key personnel/billets during vehicle mounted operations include—

- Mission commander.
- Vehicle commander.
- Driver.
- Navigator.
- Crew-served or primary weapon gunner.
- Security or dismount.

The squad will typically operate with three to four Marines in each vehicle, meaning that certain personnel will fill multiple roles. For example, the mission commander (i.e., squad leader) typically acts as one of the vehicle commanders. The vehicle commanders (i.e., fire team leaders) also serve as the vehicle navigators. Should a fourth squad member be assigned to the vehicle, the individual typically assumes a security position in the rear of the vehicle, often serving as a dismount when the situation requires.

Each squad member must be able to perform any duty in the vehicle. Prolonged exposure to night vision equipment can quickly fatigue the eyes of the drivers. Harsh climates and rough terrain may impact the gunners standing in the turret. Contingencies or emergencies may require multiple positions to be filled without notice. Successful squad leaders rotate personnel to increase proficiency and maintain situational awareness.

Vehicle-mounted squads offer commanders flexible assets capable of performing a wide variety of missions and functions, such as—

- Movement of very important personnel, contacts, and high value personnel.
- Resupply operations.
- Mounted reconnaissance.
- Acting as part of a mobile assault platoon.
- Acting as a quick reaction force (QRF).
- Conducting civil-military operations.
- Performing security escort missions for enablers (e.g., explosive ordnance disposal or civil affairs teams).

Operations

Vehicle-mounted squad operations conducted across the range of military operations should have the following attributes: aggressiveness, situational awareness, unpredictability, and flexibility (see figure 7-3). Squads should use aggressive driving skills and position vehicles, weapon systems, and personnel to dominate the environment and exhibit the squad's ability to inflict lethal effects if it becomes necessary. Squads must maintain 360-degree security. Marines must become proficient in three-dimensional threat scanning (i.e., depth, width, and elevation). When conducting stability activities, this means studying the individual movement and behavior of the local populace to aid in detecting potential threats. Squads must rely on speed and agility to accomplish the mission and protect personnel and equipment. Vehicle speed should be proportionate to the terrain and mission. In general, the squad should move as quickly as possible while maintaining the safety of Marines and innocent bystanders. Drivers must become proficient at maneuvering their vehicles through or around obstacles, vehicles, and the local population. The entire convoy must be prepared to rapidly change course with well-rehearsed SOPs and battle drills.

Operational Attributes

- Aggressiveness
- Situational Awareness
- Unpredictability
- Flexibility

Figure 7-3. Mounted Operations.

Threats to Vehicle Mounted Squads. Threats to mounted squads typically fall under two broad categories according to the type of movement being conducted—off-road movements and convoy/urban movements. The vehicle-mounted detachments must recognize, plan, and rehearse actions to counter all potential threats.

Off-Road Movement. The most likely threats encountered by mounted squads conducting off-road movements are—

- Armored or mechanized forces.
- Light-skinned non-tactical vehicles (sometimes called "technicals").
- Indirect fires (i.e., artillery and mortars).
- Enemy manned and unmanned aircraft.
- Direct fires which may consist of—
    - Near and far ambushes.
    - Snipers.
    - Crew-served weapons.
    - Antiarmor weapons.
- Surveillance (i.e., enemy soldiers, adversaries, and civilian personnel).

Convoys/Urban Movements. The threats that are most likely to be encountered by vehicle-mounted squads conducting convoys and urban movements include the following:

- IEDs (i.e., including vehicle-borne IEDs).
- Suicide bombers (they may be male, female, juvenile, or elderly).
- Direct fire, including from—
    - Locations often used for ambushes, such as overpasses, alley entrances, traffic circles, and intersections.
    - Rooftops and balconies.
    - Checkpoints and entrances to restricted locations (e.g., police stations or military complexes) in attempts to provoke fratricide.
- Hostile crowds.

Mounted Movement and Formations. During mounted operations, combat formations apply to movement and halts from squad to company levels. Squads typically use five formations—column/staggered column, line, wedge, vee, and left/right echelon—in much the same manner as when dismounted. The type of formation depends on the enemy threat, measure of control, and terrain. The interval between vehicles varies according to visibility, terrain, and/or weapon ranges.

When the squad conducts a halt as part of a larger unit movement (i.e., platoon or company), it may employ the coil or herringbone formation. When operating independently (e.g., in a four vehicle configuration), it may employ a formation resembling a patrol base or "Y" formation, wherein each fire team assumes a position from 12 to 4 o'clock, 4 to 8 o'clock, or 8 to 12 o'clock, leaving the squad leader's (i.e., mission commander's) vehicle in the center (see figure 7-4).

For halts lasting more than five minutes, the squad leader should deploy dismounted security to sweep the area to negate threats from mines, IEDs, and enemy dismounted infantry. The longer the halt, the more formal and robust the defensive posture should be, including such measures as occupying high ground, establishing overwatch or LP/OPs, and establishing a local patrol plan.

Off-Road Movement. Traveling off-road lessens the chances of enemy observation or contact. Cross-country routes often provide more cover and concealment for the squad and reduce the chance of an enemy ambush. Off-road travel often involves a slower rate of movement. Terrain such as swamps, wooded areas, or jungles may be so rough that vehicles can traverse it only at extremely low speeds. Disadvantages of off-road movement include leaving more noticeable vehicle tracks and signs of passage, a greater chance of tire failure and vehicle stress, and complex navigation. In order to identify and address potential problems, the squad must rehearse cross-country movement in terrain that resembles the area of operations.

Squads conduct operations utilizing three techniques when mounted—traveling, traveling overwatch, or bounding overwatch. Each technique is employed for specific situations and conditions (see table 7-4).

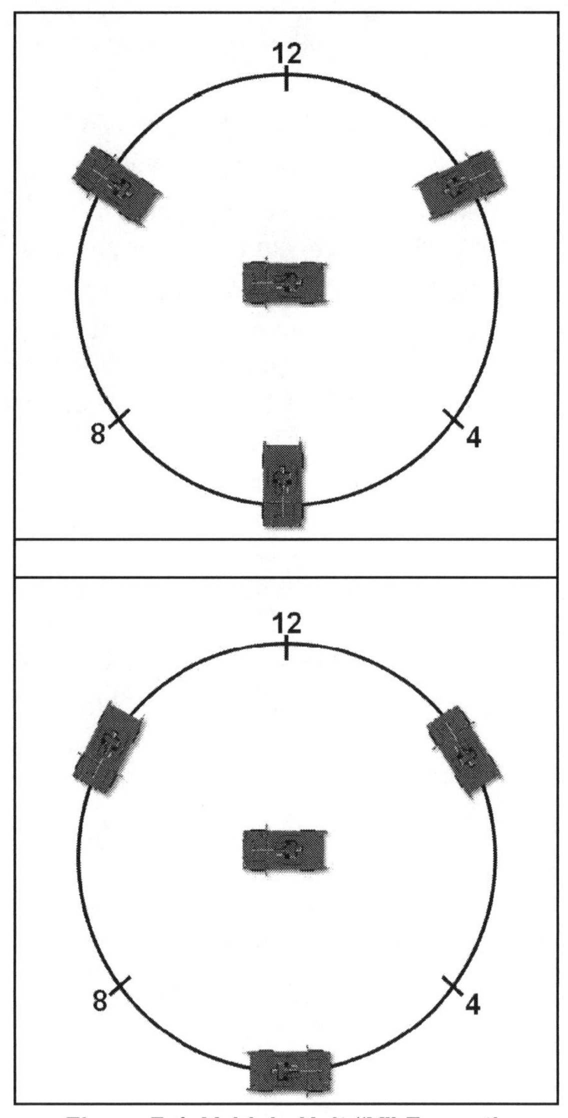

Figure 7-4. Vehicle Halt "Y" Formations.

Table 7- 4. Movement Techniques and Characteristics.

Movement Techniques	When Typically Used	Characteristics			
		Control	Dispersion	Speed	Security
Traveling	Contact unlikely	More	Less	Fastest	Least
Traveling Overwatch	Contact possible	Less	More	Slower	More
Bounding Overwatch	Contact expected	Most	Most	Slowest	Most

Movement Considerations. The considerations for mounted movement are similar to those for foot-mobile movement in many ways. However, squad leaders must remember that speed is often relative, and that deploying dismounted security often plays an integral role in mounted movement. Some situations that dismounted personnel play an integral part in include:

- Movement across open areas. Dismounted security moves rapidly across the open area to reconnoiter and secure the far side. On signal, the remainder of the squad (i.e., mounted) crosses the danger area without stopping.
- Movement through wooded areas. When moving through wooded areas, the squad leader should deploy dismounted security to move to the front and on the flanks of vehicles.
- Movement through defiles. A defile is a narrow gorge, pass, or similar man-made feature that restricts lateral movement. If enemy contact is imminent or probable and bypassing the defile is not possible, then dismounted security should clear the defile prior to the passage of vehicles.
- Movement through built-up areas. If the squad must move through a built-up area with an unknown enemy situation, the squad leader deploys dismounted security that clears the areas in advance of the vehicles. Squad leaders should consider establishing dismounted overwatch positions along the route, paying particular attention to upper floor windows and openings (for more information, see chapter 6).

Immediate Action and Battle Drills

The worst-case scenario is for the squad to encounter the enemy at a time and place that is most advantageous to the enemy. Carefully created SOPs and well-rehearsed immediate action and battle drills enable the squad to respond rapidly to such situations. Regardless of precautions taken, the squad must be prepared to make contact with the enemy by—

- Keeping weapons systems manned.
- Maintaining proper vehicle intervals.
- Enforcing movement discipline.

Should contact with the enemy be unavoidable, the squad should make contact using the smallest possible element (e.g., a single vehicle). This action allows the remaining vehicles to fire and move in support of the engaged vehicle.

During unexpected enemy contact, the squad should seek to break contact and place as much distance between it and the enemy as terrain and visibility allow. Squad and unit SOPs, as well as experience, help to shape immediate action and battle drills. In general, the most effective technique used to break contact is usually bounding away from the enemy in pairs.

Immediate Action Drills. Mounted immediate action drills are usually similar to dismounted drills (see chapter 5). Whether the contact is from the front, rear, flanks, near or far ambushes, or in open or restrictive terrain, the goal of the squad should be to conduct the following procedures:

- Identify the source of the contact by calling out, *Contact front/right/ left/rear.*
- Return fire while moving, if possible.
- Obscure the kill zone with smoke and employ fragmentation grenades.
- Establish fire superiority.
- Use an appropriate technique to withdraw or assault through the kill zone.
- Maintain support by fire and overwatch throughout maneuver.
- Deploy dismounted Marines to evacuate vehicles if rendered unrecoverable.

Battle Drills. Battle drills exist at the vehicle and crew level, and represent those basic actions needed for the vehicle to perform as an effective part of the unit in unexpected circumstances. Mounted battle drills include the following:

- Dismount drills (both rapid and normal).
- Disabled vehicle or rollover.
- Disabled weapons and reload drills.
- Incapacitated driver.
- Establishing hasty roadblocks or traffic control points.
- Sensitive equipment destruction plan.
- Vehicle recovery plan.

MECHANIZED OPERATIONS

Within the Marine Corps, mechanized forces are task-organized within the structure of the MAGTF. Squads participate as part of a mechanized and tank company team ground maneuver element that typically attacks as part of a larger mechanized force, such as a battalion or regimental-sized task force. The company team can be used to—

- Support the movement of another unit by fire.
- Serve as a maneuver element.
- Operate in reserve.

To best exploit the mechanized force's offensive capabilities, infantry, tanks, and AAVs must work together in pursuit of a common goal. Each element of the mechanized force provides a degree of mutual support to the other elements. Assault amphibian units and tank units support the infantry (squad) by—

- Providing mobile protected firepower.
- Neutralizing or destroying hostile weapons by fire and movement.
- Breaching lanes through wire for dismounted infantry.

- Neutralizing fortified positions with direct fire.
- Supporting dismounted infantry by direct fire.
- Providing protection against long-range antiarmor fires.
- Leading the attack whenever possible.
- Assisting in the consolidation of the objective.

The infantry squad supports AAV and tanks by—

- Breaching/removing non-explosive antiarmor obstacles.
- Assisting in neutralizing or destroying enemy antiarmor weapons.
- Designating targets for tanks or AAVs.
- Protecting tanks and AAVs from enemy infantry and antiarmor weapons.
- Leading the attack dismounted when necessary.
- Clearing restrictive terrain such as urban, swamp, mountainous, or wooded areas.
- Conducting dismounted security patrols.

Maneuver Considerations

During mechanized attacks, speed is essential, and should be maintained to the greatest degree possible. The determination of whether the squad will be mounted or dismounted is based on METT-T analysis. If the squad remains mounted, then speed may be maintained. If it is dismounted, then speed is reduced. The following considerations should be used.

Tanks Lead. When tanks lead, they maneuver together with mechanized infantry, with the rifle squads mounted in AAVs in trace, supported by the base of fire element and available supporting arms. Typically, the squad remains mounted when—

- Enemy antiarmor fires can be effectively bypassed or suppressed.
- Terrain is relatively open or obstacles can be easily overcome.
- Environmental conditions offer good trafficability and visibility.

Infantry Mounted. Mobility and limited armor protection allow the squad to cross the battlefield quickly when they are mounted. The squad typically remains mounted when—

- Enemy resistance is extremely light.
- Enemy personnel are in hasty positions.
- Suppressive fires have reduced enemy antiarmor fires.
- Terrain allows rapid movement onto and across the objectives.

Infantry Dismounted. Dismounted squads provide cover to the flanks and rear of the mechanized force by employing organic fires, directing fires from the base of fire element, and providing supporting arms against enemy positions. The squad may move far enough behind tanks to avoid being hit by enemy fire directed at the tanks. However, on occasion when tanks are "buttoned up," such as in urban terrain, the squad leader may utilize a tank's infantry phone to designate targets and communicate with the tank's crew. This technique permits close coordination and maximum mutual support, but sacrifices the speed and mobility of the AAVs and tanks. Infantry leads dismounted when—

- Terrain and vegetation are restrictive, canalizing movement into likely ambush sites or minefields.
- Visibility is limited.
- Antiarmor fires cannot be bypassed or suppressed.
- Significant obstacles or fortifications are encountered that prevent mounted movement.

Amphibious Assault Vehicle
This section describes the capabilities, limitations, and considerations for employing AAVs in mechanized operations.

Capabilities. The rifle squad leader's understanding of the capabilities and limitations of the AAV are important in maximizing its operational utility on the battlefield. The AAV is capable of open ocean operations from offshore shipping through rough seas and plunging surf. Without modification, it is capable of traversing beaches, crossing rough terrain, and performing high speed operations on improved roads. The AAV provides the squad with armor protection as well as mobility on land and water.

Protection. The AAV provides armor-protected mobility to a rifle squad, which can dismount to carry the fight to the enemy while the AAV crew fights from overwatch. The hull of the AAV is constructed from welded plates of ballistic aluminum, providing a high degree of protection against small arms fire up to .30 caliber at 300 meters, as well as 105-mm high explosive air burst fragmentation at 15 meters.

Maneuver. The AAV is capable of worldwide operation in nearly any terrain, on land or water. Due to its low ground pressure, the AAV is capable of operating in soft soil that is inaccessible to M1A1 tanks or light armored vehicles. Each AAV (i.e., personnel variant) carries 3 crew members and up to 21 embarked Marines, depending upon internal cargo requirements. The AAV has the following capabilities:

- On land:
  - 300-mile operating range.
  - Cruising speed of 25 miles per hour on flat, hard surface roads, with a maximum land speed of 45 miles per hour.
  - Can operate on forward slopes of 60 percent and side slopes of 40 percent.
  - Can cross 8-foot trenches and 3-foot vertical obstacles.
- On water:
  - Capable of operating in calm to moderate sea conditions.
  - Capable of negotiating up to 10-foot plunging surf and self-righting from a 180-degree roll.
  - Maximum water speed of 8.2 miles per hour.
  - Waterborne range of 45 miles in calm seas.
  - Capable of long-distance water marches, limited only by sea conditions and the effects of motion on embarked personnel.

Firepower. See table 7-3.

Operational Considerations

Safety. The combination of heavy equipment, high mobility, limited observation, and waterborne maneuvers creates conditions that require strict adherence to instructions and SOPs concerning mechanized operations. Carelessness, recklessness, shortcuts, and inattention can result in serious injuries or death to rifle squad members and crew, as well as damage to equipment. Squad members must be instructed, inspected, reminded, and corrected on a continuous basis about following proper procedures. Many mishaps can be directly attributed to not following proper procedures and safety guidelines.

Though safety is the responsibility of all Marines; leaders embarked on AAVs and mechanized unit leaders share specific responsibilities to ensure safe operations. While embarked aboard AAVs, squad leaders are responsible for briefing their Marines on their responsibilities as passengers.

While embarked, Marines should pay strict adherence to the following safety guidelines:

- No smoking should be allowed on or in AAVs.
- Marines must wear helmets (at minimum) and not ride topside.
- Dismounted Marines must give vehicles a wide berth due to the limited fields of vision of "buttoned up" AAV crews.
- Open hatches must be secured with restraints while moving; hatches can slam shut and cause serious injuries to embarked personnel.
- Marines should not ride with chests extended above hatch openings.
- Ground guides must be used while backing or moving vehicles, specifically during periods of limited visibility.
- Marines must stand clear when the ramp is being raised or lowered.

Prior to waterborne operations, squad leaders are responsible for ensuring that all embarked Marines have a clear knowledge and understanding of what their safety roles and responsibilities are when on AAVs in the water. Before the launch order to enter the water, Marines must be properly outfitted with their lifejackets, helmets, body armor, and weapons.

Personal equipment for waterborne operations should adhere to the following:

- Lifejackets must receive an operational check before being worn, to include—
  - A serviceable carbon dioxide cartridge.
  - Light or chem light.
  - Whistle.
  - Three puffs of air in jacket.
- Helmets must be worn at all times with chin straps fastened.

- Life jackets must be worn under body armor and fighting loads, and worn loosely enough to jettison them immediately if required.
- Personal weapons should be in condition three, with muzzles inverted during transit to shore. Weapons should be changed to condition one upon exiting vehicles.

All attempts should be made to limit the time Marines are waterborne because of the combined effects of heat, noise, fumes, seasickness, and claustrophobia. Ventilation in the troop compartment may help to alleviate discomfort and sickness.

Safety is an important consideration on land, as well. Amphibious assault vehicles are not air conditioned, and can cause heat injury to embarked Marines if the right precautions are not followed. Squad leaders need to be aware of the hazards that may occur in the environment in which the AAVs are operating in, and have the ability to adjust. Operations over land can be influenced by unfamiliar climates and terrain.

Because heat injuries are probable in closed vehicles, Marines must have proper ventilation and constant hydration in hot weather. Water cans, canteens, and bladders should be filled at every opportunity; leaders must ensure that Marines are hydrating at all times and be watchful for signs of distress. Marines should carry at least two empty one-liter (at minimum) water bottles or equivalent in order to relieve themselves during mounted movement.

Besides hot weather, high dust conditions can reduce visibility for AAV crews. Marines but be extra vigilant in dusty conditions, especially if they must operate dismounted. Marines must know and understand the basic operational procedures of the AAV's up-gunned weapons station to prevent possible fratricide.

The AAVs must be properly ventilated during cold weather conditions due to the risks posed by vehicle heaters and the danger of carbon monoxide poisoning to embarked squads. Marines working around all metal vehicles run the risk of contact frost bite and hypothermia if not properly equipped. Squad leaders must ensure that Marines are wearing the proper protective equipment to mitigate these risks. The AAV may have a tendency to slide on icy surfaces, which can cause great risks to dismounted squads.

Squad leaders must be aware of safety concerns related to weather, CBRN operations, lasers, breaching using the AAV-mounted mine clearing linear charge (i.e., MK-154), abandon vehicle procedures (i.e., waterborne), and onboard fires. Many of the weather concerns are the same as when dismounted (e.g., lightening, high winds, heavy rains); the primary concern is what effect the conditions have on the AAV's capabilities.

Squad leaders must ensure that all Marines know and understand all SOPs pertaining to AAV operations, and their associated requirements. The AAV has unique safety guidelines and procedures that must be reviewed. For more information on AAV procedures, see MCTP 3-10C, *Employment of Amphibious Assault Vehicles*.

Urban Operations. Though AAVs have the same mobility restrictions as tanks in built-up areas, they can provide valuable combat support and/or combat service support roles for squads in urban terrain. The AAV's mobility can support the squad with the following:

- Rapid movement across open areas during the seizure of a foothold when covered and concealed routes are not available.
- Direct fire from an overwatch position during the seizure of a foothold.
- Acting as a mobile command echelon.
- Providing rapid movement in cleared areas for reserve forces.
- Serving as evacuation platforms for detainees, civilians, or CASEVAC.
- Providing concentrated sustained direct fires in a combat support role.
- Clearing lanes of suspected or confirmed explosive obstacles with the MK154.

Equipment. When conducting mounted operations, squad leaders must pay special attention to their squads' individual and special equipment. Due to the lack of internal AAV storage, the squads' equipment is often stowed externally. Squad leaders should consider the following:

- All special equipment should be carried in assault loads and stowed internally.
- When conducting waterborne operations, external loads need to be thoroughly waterproofed.
- Loads should be strapped securely with shoulder straps and then "dummy corded" with safety lines using lengths of rope and carabiners.
- In urban operations, external equipment needs to be mounted securely and as high as possible to prevent locals from taking them.

Tank/Infantry Integration

Using tanks in mounted operations maximizes the ground mobility, protection, shock action, and firepower of these combat vehicles to destroy the enemy's will to resist. Combat power is generated through the mass employment of tanks and enhancing the infantry's mobility by mounting them on supporting vehicles. Refer to MCTP 3-10B, *Marine Corps Tank Employment*, for more information on tank operations and their related infantry and safety precautions.

The use of cross-attachment at the company level results in the creation of tank and mechanized teams. While numerous variations are possible, such as a dismounted squad receiving tank support or a tank receiving a mounted infantry squad, the general principles of tanks and infantry working together remain the same.

To exploit the mounted force's capabilities, tanks and mounted infantry squads work together to achieve a common goal. Each element of the mounted force provides a degree of support to the other elements.

Tanks support mounted squads by—

- Providing mobile protected firepower.
- Neutralizing or destroying hostile weapons by fire and movement.
- Clearing paths for dismounted infantry through obstacles.

- Neutralizing fortified positions with direct fire.
- Supporting dismounted infantry by direct fire.
- Assisting in the consolidation of the objective.

Rifle squads can provide support to tanks by—

- Breaching or removing antiarmor obstacles.
- Neutralizing or destroying enemy antiarmor weapons.
- Designating targets for tanks in built-up urban terrain.
- Protecting tanks from enemy infantry.
- Clearing bridges and fording areas.
- Clearing restrictive terrain, such as urban, swamp, or heavily wooded areas.
- Conducting dismounted security patrols.

Employment. Based on METT-T and the tactical situation, there are usually three methods to employ tanks and mounted squads together:

- Tanks and mounted squad together—
    - Exploits the mobility, speed, armor-protected firepower, and shock action of the mounted force.
    - Disorganizes the enemy's defense by using tanks to breach obstacles prior to the infantry dismounting.
    - Conserves the energy of the infantry by reducing the distance to travel on foot, and reduces the amount of exposure of the infantry to enemy fires.
- A mounted squad supported by fire from tanks and AAVs—
    - Increases vehicle survivability while still employing the positive characteristics of the tank and AAV weapon systems.
    - Enables vehicles to bound forward and maintain effective support after the squad clears areas during its advance.
    - Increases the survivability of the infantry in the face of significant antiarmor threats.
- Multi-axis (i.e., a combination of the two).

Transporting Infantry. On very rare occasions, squads may be required to be transported on a tank. This is usually required only when contact is not expected. Squad leaders should consult the tank commander on positioning Marines on tanks (see figure 7-5).

Riding on the outside of the vehicles is hazardous. Therefore, the squad should only ride on tanks when the need for speed is great and more suitable options are not available. Tanks can also react more quickly without Marines riding on top. Marines on the outside tanks are vulnerable to—

- The effects of all types of fire (i.e., both direct and indirect).
- Obstacles that may cause tanks to turn suddenly or knock them off the tank (e.g., tree limbs), including the traversing of the turret gun.

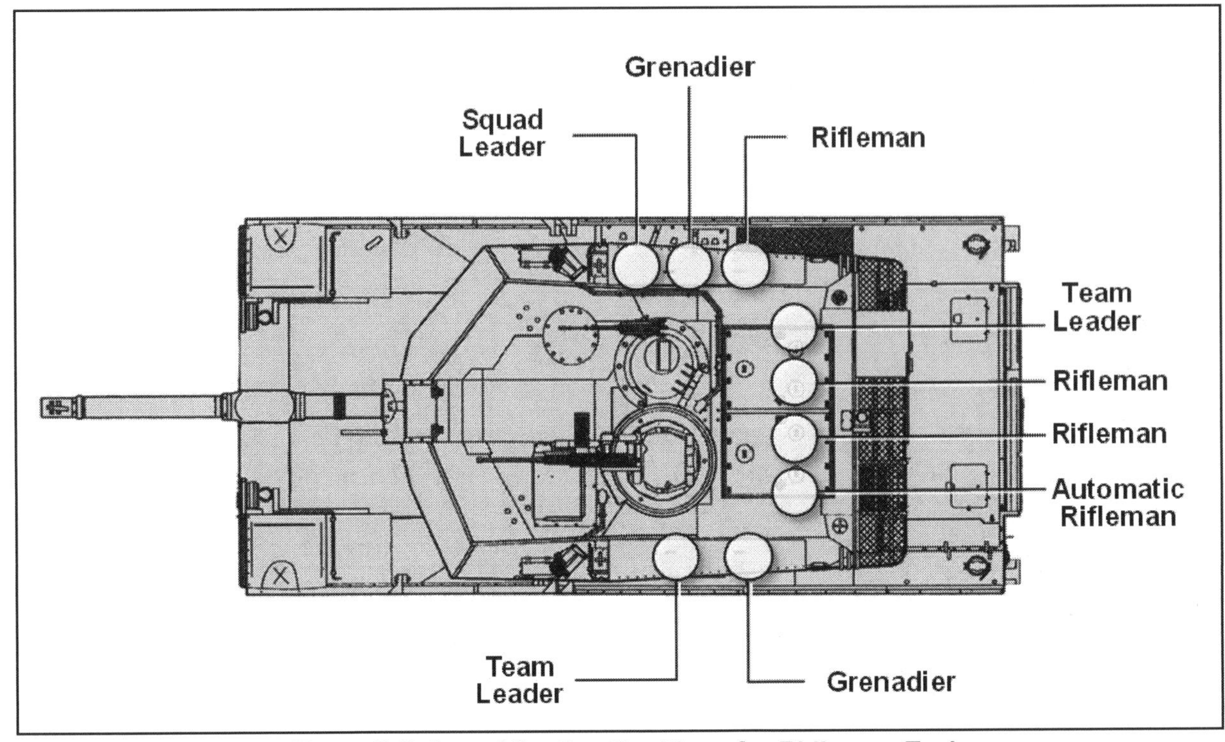

Figure 7-5. Squad Seating Positions for Riding on Tanks.

The only advantages the squad gains from riding on tanks are increased speed of movement and increased haul capability. In this case, the following apply:

- Marines should avoid riding on the lead vehicle of a section or platoon.
- The squad's leaders should be positioned with the tanks' vehicle leaders.
- Contingency plans for chance contact or danger areas should be discussed and prepared for.
- The rifle squad should dismount and clear chokepoints or other danger areas.
- Air guards and sectors of responsibility for observation should be assigned.
- Leaders should ensure all Marines remain alert and stay prepared to dismount immediately.
- In the event of contact, the tank will immediately react as required for its own protection.
- The squad is responsible for its own safety; rehearsing rapid dismount drills is a must.
- The M1 series tank is not designed to carry Marines easily. Marines must not move to the rear deck. Engine operating temperatures make this area unsafe for riders.
- One squad can ride on the turret. Marines must mount in such a way that their legs cannot become entangled between the turret and the hull by an unexpected turret movement. Rope may be used as a field expedient rail to provide secure handholds.
- Marines must ride to the rear of the smoke grenade launchers; this should keep personnel clear of the coaxial machine gun and laser range finder.
- The squad must always be prepared for sudden turret movement. Leaders should caution Marines about sitting on the turret blowout panels. If there is an explosion in the ammunition rack, the panels blow outward to lessen the blast effect in the crew compartment.
- If enemy contact is made, the tank should stop in a covered and concealed position and allow squad time to dismount and move away from it. This action needs to be rehearsed before movement.

- The squad should not ride with anything more than their fighting loads; excess personal equipment should be transported elsewhere.

Combat Vehicles and Infantry Squad Formations. The elements of METT-T assist the squad leader in selecting formations for combat vehicles and dismounted infantry. The same principles for selecting formations with dismounted squads apply for combat vehicles moving with dismounted Marines. The squad leader can employ the fundamental vee and wedge formations for combat vehicles to meet the needs of the mission. After squad leaders combine tank and infantry elements into one combat formation, it is their responsibility to ensure that proper communication and fire control measures are employed to maximize lethality and prevent fratricide.

After selecting the combat formations for tanks and dismounted infantry, the squad leader decides whether to lead with tanks, infantry, or a combination of the two. Depending on the tactical situation, the default technique is to lead with the dismounted squad.

Leading with Dismounted Squad. The dismounted squad is far better suited for leading under the following conditions (see figure 7-6):

- A route leads through restrictive urban or rural terrain.
- Stealth is desired.
- Enemy antitank minefields are suspected.
- Enemy antitank teams are anticipated.

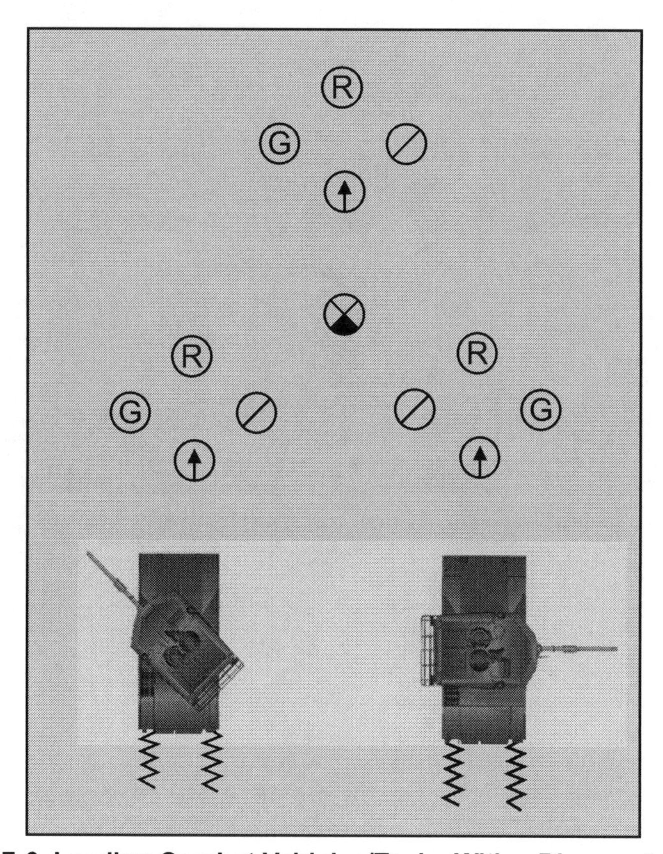

Figure 7-6. Leading Combat Vehicles/Tanks With a Dismounted Squad.

Leading with Tanks. A squad leader might decide to lead with tanks during the following situations (see figure 7-7):

- There is an armored or tank threat.
- Moving through open terrain with limited cover or concealment.
- There is a confirmed enemy location/direction.
- There are suspected enemy antipersonnel minefields.

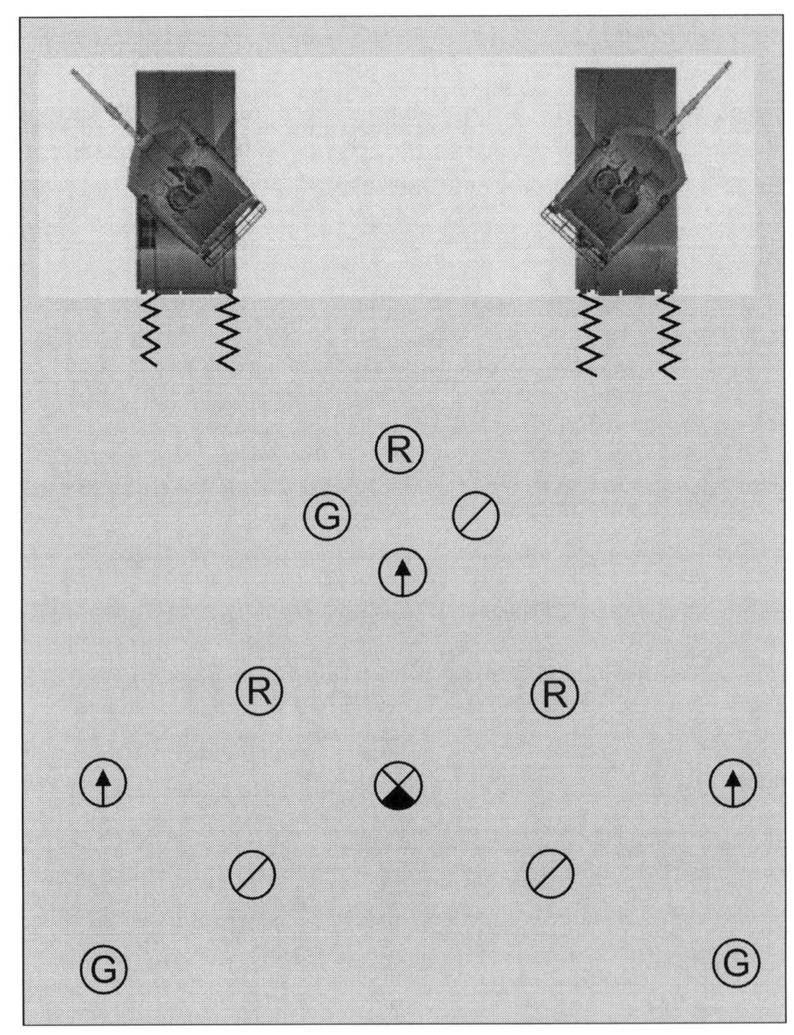

Figure 7-7. Leading a Dismounted Squad With Combat Vehicles/Tanks.

Leading with Both Combat Vehicles/Tanks and Dismounted Squad. The squad leader may choose to centrally locate with the combat vehicles/tanks in their formations (see figure 7-8) when —

- Flexibility is desired.
- The enemy location is unknown.
- There is a high threat of dismounted enemy antitank teams.
- The ability to mass the fires of the combat vehicles quickly in all directions is desired.

> *Note:* Tanks' main guns fire high-velocity, armor-piercing, discarding sabot rounds that pose hazards to dismounted infantry. Dismounted Marines should be at 300 meters to the left or right of the line of fire and at least 1,300 meters to the front of a tank firing its main gun (see figure 7-9). Any Marines within this danger area must have adequate cover.

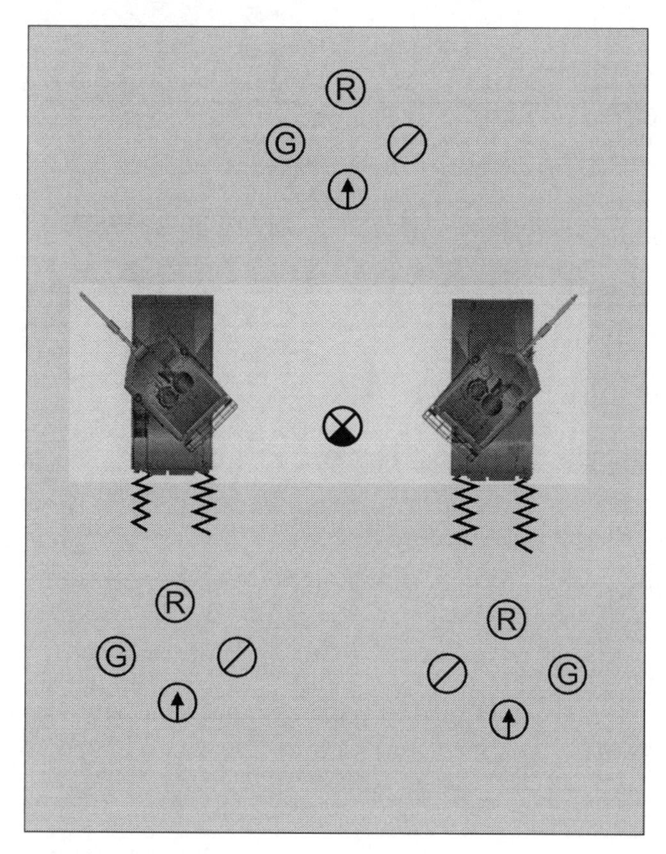

Figure 7-8. Squad Centrally Located With Combat Vehicles/Tanks.

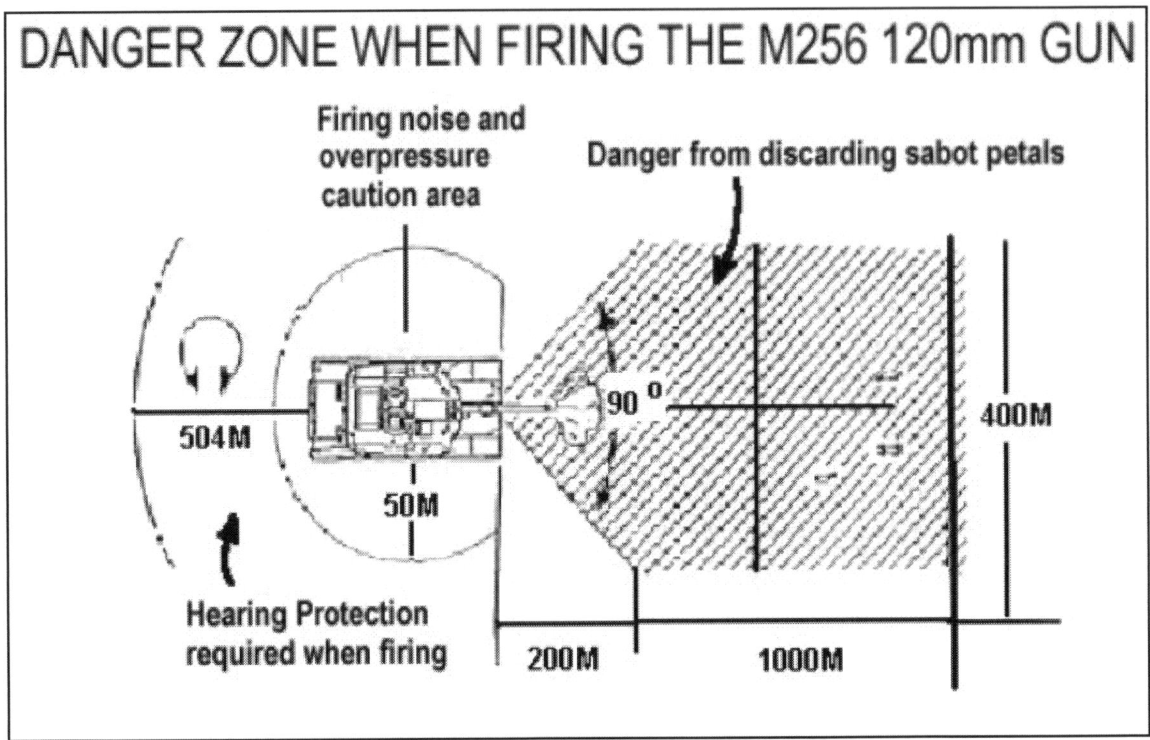

Figure 7-9. Tank Main Gun Danger Zones.

AIR ASSAULT OPERATIONS

Air assaults are high-risk, high-payoff missions. When properly planned and vigorously executed, these missions allow leaders to rapidly generate and apply combat power to areas that were not previously occupied. An air assault can provide leaders the means to control the tempo of operations, enabling rapid execution to retain or exploit the initiative. For further information on planning and conducting air assaults, refer to MCTP 3-01B, *Air Assault Operations*.

Air assaults should never be confused with air movements, as they are separate and distinct. Air movement is the transportation of units, personnel, supplies and equipment, to include air drops and air landings. Air movement usually relies on utility and cargo aviation assets.

Capabilities and Limitations

Air assault operations are most effective in environments where limited lines of communication are available to the enemy, and the enemy lacks air superiority and effective air defense systems. They should not be employed for deliberate operations over an extended period, and they are best employed in situations providing a calculated advantage due to surprise, terrain, threat, or mobility. In particular, air assault operations are employed in missions requiring—

- Massing or shifting combat power quickly.
- Using surprise.
- Using flexibility, mobility, and speed.
- Gaining and maintaining the initiative.

Capabilities. Air assault operations can extend the battlefield, providing the ability to move and rapidly concentrate combat power. Table 7-5 provides a list of capabilities of the current family of Marine Corps air assault assets. Air assault operations may also provide the following capabilities:

- To attack enemy positions from any direction.
- To conduct attacks and raids within the operational area.
- To overfly and bypass enemy positions and obstacles, and to strike objectives in otherwise inaccessible areas.
- To rapidly commit reserves into action.
- To conduct and support deception.
- To rapidly defend and secure key terrain or key objectives.
- To delay larger forces without becoming decisively engaged.

Table 7- 5. Aircraft Characteristics.

	MV-22	CH-53E	AH-1Z	UH-1Y
Mission	Assault transport of heavy weapons, equip., and supplies.	Assault transport of heavy weapons, equip., and supplies	Fire support	Utility
Alt. Mission	Assault troop transport, casualty evacuation, TRAP	Assault troop transport, casualty evacuation, TRAP	FAC(A), TAC(A), escort, aerial reconnaissance	FAC(A), TAC(A), escort, aerial reconnaissance
Crew Configuration	2 pilots 1 crew chief/gunner 1 aerial observer/gunner	2 pilots 1 crew chief/gunner 1 aerial observer/gunner	2 pilots	2 pilots 1 crew chief/gunner 1 aerial observer/ gunner
Max. Speed	316 mph	173 mph	255 mph	189 mph
Max. Endurance	1,011 miles	621 miles	426 miles	300 miles
Aerial Refuel Capable	Yes	Yes	No	No
Weapons Systems	1 x M240 or .50 cal machine gun on ramp GAU-17 mini-gun belly mounted (nonstandard)	2 x .50 cal XM 218 Machine Guns	1 x 20-mm Gatling gun 19 x 2.75 inch rockets 2 x Sidewinder air-to-air missiles 16 x Hellfire missiles	2 x External rocket stations; 2 x Mounts for M240 or .50 cal machine gun or 7.62-mm GAU-17/A Gatling guns
Lift Capacity	24 Marines seated 32 "floor loaded" 20,000 lbs internal 15,000 lbs external	37 to 55 passengers 24 Marines at 250 lbs each (6,000 lbs w/ 4+00 hour endurance Cargo – 32,000 lbs	No cargo/ passenger ability	6,600 lbs Up to 10 passengers or 6 litters
LEGEND: FAC(A) forward air controller (airborne) TAC(A) tactical air controller (airborne)				

Limitations. Air assault operations rely heavily on the availability of aviation assets to support continued operations. As such, operations may be limited by—

- Adverse weather, extreme heat or cold, and other environmental conditions (e.g. blowing sand or snow) which limit flying time, aircraft lifting capability, or altitude and elevation restrictions that can affect operational capabilities.
- Reliance on air lines of communication.
- The availability of suitable LZs and pickup zones (PZs) due to heavily wooded, mountainous, urban, jungle, or other complex terrain.
- Battlefield obscuration that can limit aircraft flight.
- The availability of organic fires, sustainment assets, and protection.
- High fuel and ammunition consumption rates.

Planning

Planning for air assault operations mirrors the squad leader's troop leading step process; incorporating parallel and coordinated planning necessary for successful execution. Standardizing operations between elements conducting the air assault significantly enhances the ability of the squad to accomplish the mission.

Air assault planning is as detailed as time permits, and includes completing written orders and plans. Within time constraints, the squad leader should carefully evaluate the capabilities and limitations of the total force and develop a plan that ensures a high probability of success. The planning time should abide by the one-third/two-thirds rule described in chapter 2 to ensure fire team and other element leaders shave enough time to prepare, plan, and rehearse.

Squads typically participate in air assault operations as part of a larger force. However, there may be situations when a squad is tasked to be the maneuver element for a tactical recovery of aircraft and personnel (TRAP) force or as a QRF. In these instances, the squad leader is responsible for developing detailed plans in coordination with various other elements. Air assault mission require the development of five basic plans: the ground tactical plan, the landing plan, the air movement plan, the loading plan, and the staging plan (see figure 7-10).

Air Assault Plans

Ground Tactical Plan
Landing Plan
Air Movement Plan
Loading Plan
Staging Plan

Figure 7-10. Five Plans of Assault Operation.

Air Assault Plans

Ground Tactical Plan. The ground tactical plan is the squad leader's vision of how the squad will accomplish its assigned mission, including the scheme of maneuver and fire support plan.

Scheme of Maneuver. The scheme of maneuver should include the—

- Mission statement.
- Objectives.
- LZs.

- Total weight to be lifted.
- Number of personnel to be transported.
- Forms of maneuver to be employed.
- Distribution of the squad.
- Detailed timelines, including anticipated enemy effects.

Fire Support Plan. The fire support plan should include—

- How the scheme of maneuver is supported.
- The preparation of the LZs.
- The integration of information related capabilities from IO.

Loading Plan. The loading plan is designed to establish, organize, and control activities in the PZ and plan for the movement of the squad and its equipment to the PZ. During TRAP or QRF missions, the squad may often be required to stage at the PZ. However, the requirement for written loading instructions can be minimized by advanced planning and detailed unit SOPs.

Aircraft loads are also prioritized to establish a bump plan. A bump plan ensures that essential Marines and equipment are loaded ahead of less critical loads in case of aircraft breakdown or other problems. If all squad members and special equipment cannot be lifted, individuals must know who is to load or offload and in what sequence. Table 7-6 gives an example of an individual bump plan. Special equipment that is essential for mission accomplishment should NOT be packed into individual Marines' assault packs in case of the bump plan must be executed.

Table 7- 6. Example Squad Bump Plan.

Aircraft	Unload	Sequence
102-1	1st	Johnson
	2nd	Roberts
	3rd	Smith
102-2	1st	Jones
	2nd	Griffith
	3rd	Hodge

Air Movement Plan. The air movement plan provides for the control and protection of the air assault force during the air movement. The air movement plan includes the selection of approach and retirement lanes, control points, en route fires to suppress enemy air defenses, and the provisions for escort aircraft or other aviation. Though the squad leader should provide input, it is usually the air mission commander's responsibility.

Landing Plan. The landing plan typically consists of the squad leader's guidance concerning the desired time, place, and sequence of unit arrival. The landing plan must support the ground tactical plan. During TRAP or QRF missions, the LZ is often not known, and the squad must gain as much information about the LZ as possible while en route.

Staging Plan. The staging plan is based on the loading plan and prescribes the arrival time of the squad (i.e., personnel, equipment, and supplies) at the PZ and their proper order for movement. Loads must be ready before aircraft arrive at the PZ. The squad may be required to stage in the PZ for an extended period of time in support of ongoing missions.

Terminology

Marshalling Area. In air assault operations, this is the area where personnel are consolidated and accounted for, similar to an assembly area.

Marshalling Area Control Officer. The marshalling area control officer (MACO) is generally the platoon sergeant or platoon guide, who is responsible for accountability of all personnel entering or exiting the marshalling area. In the event of independent squad air assault operations, the MACO is typically the first fire team leader or another individual designated by the squad leader.

MACO Gate. Generally, there should only be one entrance and exit point to the marshalling area in order to simplify the accountability of all personnel entering or exiting it. This point is known as the MACO gate.

Serial. This generally consists of a fire team (i.e., four personnel), and is the smallest subdivision of personnel to be embarked. Each serial is assigned a number. They are generally organized so that if the bump plan is executed and personnel must embark on a different type of aircraft (i.e. with a different load capacity), the serials can be easily reassigned to facilitate the transportation of all personnel.

Stick. This refers to the group of personnel on an aircraft. It typically consists of multiple serials.

Task Organization

Air assault platoons are generally smaller and task-organized differently than typical infantry platoons. In turn, each squad may consist of only two fire teams. This allows the squad leader to assign an assault team and a support team.

Depending on the mission, the squad should be task-organized with an assault, support, and security element. In the case of a TRAP mission or a QRF mission in which linkup and recovery of US personnel is required, this will slightly change to search, support, and security. The security element is responsible for the seizure and security of the LZ. It holds the LZ for the duration of the mission unless it becomes unmanageable (criteria that are addressed prior to execution of the mission). The search element acts as the primary linkup and recovery element, while the support element provides internal security for the search element.

Considerations

The following subparagraphs address several considerations and procedures specific to air assault operations that a squad leader should be familiar with.

Linkup Procedures. Most air assault missions involve a linkup with other friendly forces. The linkup plan must be discussed and planned at the platoon level or higher. However, it is the squad leader's responsibility to ensure that the linkup procedures for each specific mission are known and understood by every Marine.

Mission Essential Equipment. It is the squad leader's responsibility to be familiar with all equipment required in an air assault mission, and to inspect their Marines prior to execution to ensure all equipment is on hand and operable. This includes, but is not limited to, the following:

- One white board per aircraft to communicate within the stick while airborne and provide any updates to the situation on the ground.
- One ground radio per aircraft for enhanced ability to communicate between aircraft, either by voice or digitally.
- One digital tablet per aircraft (if applicable).
- One day of supply of rations, water, and batteries per Marine (for contingency planning).
- One signal mirror per Marine. This should be included in all lost Marine plans when conducting air assault operations, as prescribed by unit SOP.
- One wag bag or empty disposable plastic bottle (one-liter minimum) per Marine for relief in case of events requiring extended time on station or staging.

Training

The following paragraphs address some specific training objectives that must be rehearsed consistently to improve chances of successful air assault operations.

Loading/Unloading Aircraft. This seemingly simple task becomes much more difficult when verbal communications are limited by the noise of the aircraft and other variables. Air crews often have different SOPs regarding when they allow passengers to unbuckle safety belts, remove life preserver units, and offload the aircraft. It is important to communicate with the air crew prior to execution in order to understand their procedures. At minimum, boarding the aircraft and fastening safety belts, as well as removing safety belts and offloading the aircraft in a tactical and efficient manner should be well rehearsed.

When loading and unloading aircraft, Marines must be aware to avoid coming into contact with the main and tail rotors. The type of aircraft (i.e., which may include other Service or coalition nation aircraft) determines which dangers may be encountered. There are typically two methods to load and unload: from the rear ramp, or from the sides. Aircraft should typically be unloaded in the reverse order of loading. Figure 7-11, figure 7-12, and figure 7-13 depict methods for unloading aircraft.

Mockups can be used to train individuals on how to approach a helicopter, how to board it, and how to unload from it. Air assault battle drills can be taught by using mockups. Therefore, much of the individual and small unit training should be accomplished using aircraft mockups. Combat support Marines can also be trained to load weapons, equipment, supplies, and ammunition on air assault aircraft by practicing on mockups.

Securing/Maintaining Security on a Landing Zone. Securing the LZ is always the first part of the ground scheme of maneuver. It is important for squad leaders to understand what their responsibility is when their aircraft touches down in a potentially hostile environment. This is developed through training and unit level SOPs.

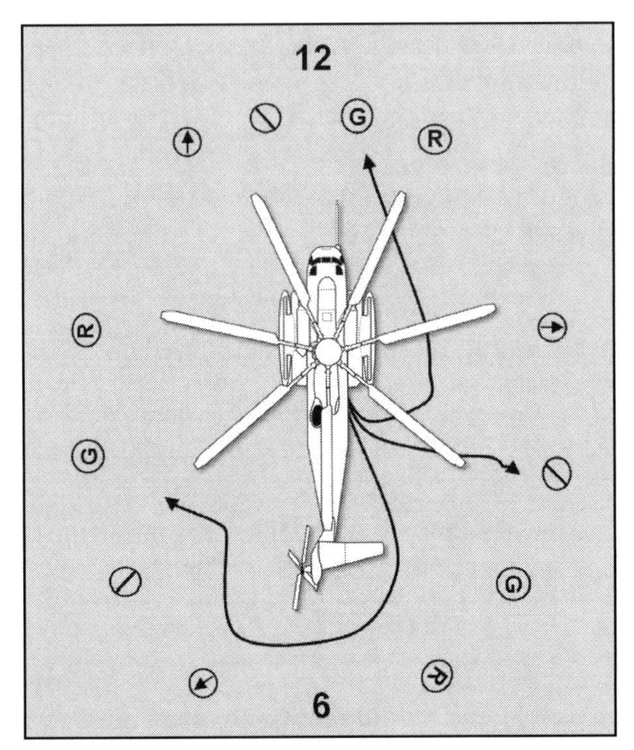

Figure 7-11. Squad Unloading From the Ramp of a CH-53.

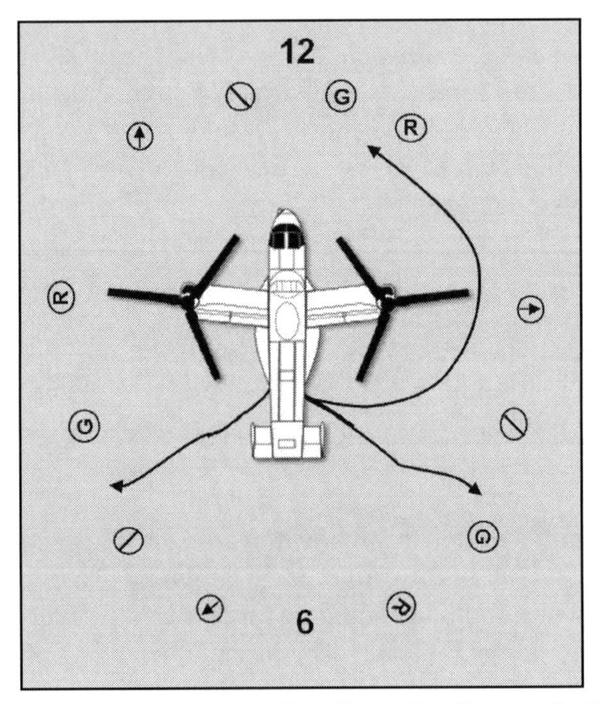

Figure 7-12. Squad Unloading From the Ramp of a MV-22.

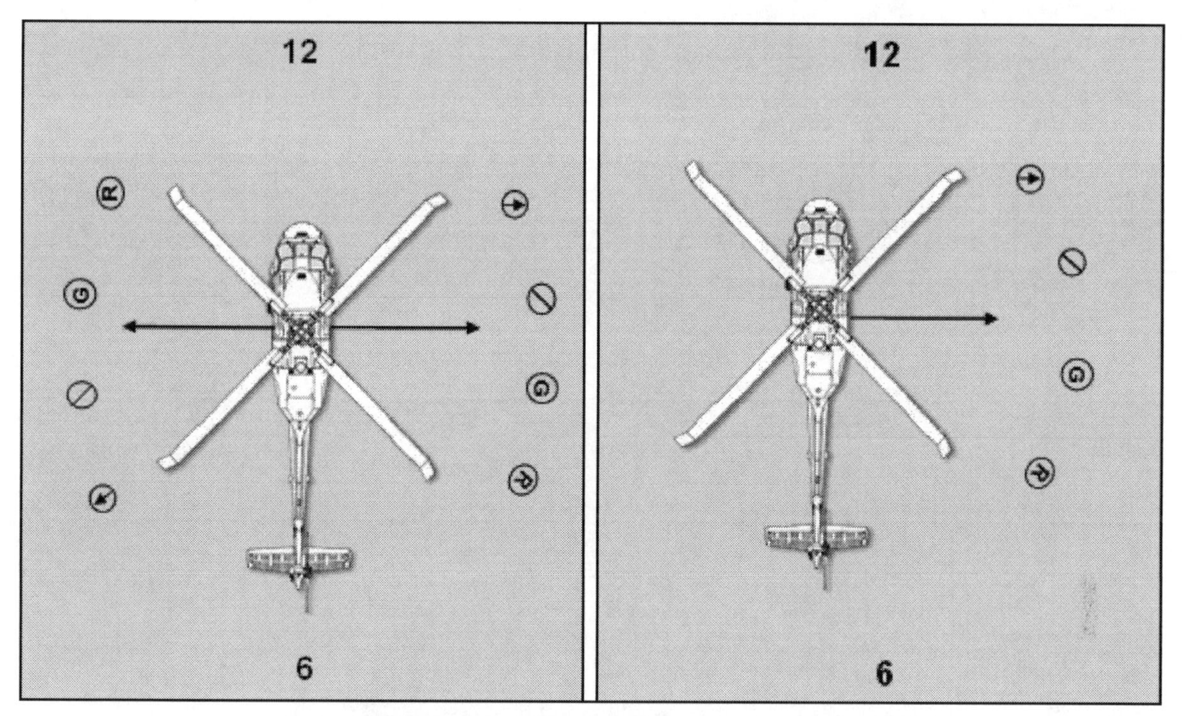

Figure 7-13. Two and One-Side Unloading From a UH-1.

MACO Procedures. Based upon the mission, it is often acceptable to conduct squad-level accountability upon landing for a mission. The squad leader is simply responsible for ensuring that all Marines are present and ready to carry out their tasks. The squad leader then leads them to the objective or linkup point. This allows for speed and decentralized execution. However, it is extremely important to conduct strict MACO procedures upon returning to the marshaling area prior to extraction, during which each person passes through the MACO gate and is accounted for by the MACO.

Warning

When exiting a CH-53, all personnel must exit to the left from the ramp. Exiting to the right may cause serious injury or death from the tail rotor.

Tablet/Digital Interoperability. Technology has advanced to the point where airborne leaders are able to communicate electronically while en route to the objective. This includes chat and the transmission of schemes of maneuver with drawings or actual operational terms and graphics on satellite images of an objective area. In order to become proficient with this technology, it is important to conduct hands-on training during tactical decision games and mission rehearsals. For more information on unit training management for rifle squads, see appendix H.

APPENDIX A
PREPARATIONS FOR COMBAT

OVERVIEW

Squad leaders can expect to receive an order from their platoon commander when planning and preparing for a combat operation. In most situations, there is time for them to prepare their own orders and issue them to their squads. Many aspects of a squad plan come from the platoon commander's order. When preparing for a mission of any type, squad leaders may be aided in their preparations by adhering to the six troop-leading steps: begin planning, arrange for reconnaissance, make reconnaissance, complete the plan, issue the order, and supervise (BAMCIS). Proper use of the troop-leading steps, in combination with pre-combat checks, inspections, and rehearsals, allow units to best prepare to accomplish a given mission.

Leaders must be certain that everyone is ready and everything is in order before a mission. If a weapon is not working, a radio's batteries are dead, or not enough water or rations are available, the mission's success and the squad's safety may be threatened. It is a small unit leader's job to ensure that their Marines have all the necessary clothing and equipment, that the equipment is in working order, that sanitary conditions are met, and that the squad can operate effectively when called upon. Squad leaders accomplish this by conducting thorough pre-combat checks and pre-combat inspections, often referred to as PCCs and PCIs. The disciplined use of pre-combat checks and inspections build unit cohesiveness and prevent complacency.

> ***Note:*** The checklists provided below are only examples. Squad leaders should utilize unit SOPs and mission-specific requirements to construct their own pre-combat check and pre-combat inspection checklists.

PRE-COMBAT CHECKS

Pre-combat checks are used to ensure that a squad is adequately prepared to execute activities and actions according to standards. They are detailed checks all units should conduct before and during all activities and actions in support of combat operations. Leaders should perform checks before each action or activity to check personnel and equipment. By requiring their squad to adhere to the established standards, a squad leader ensures that every Marine and every piece of equipment is mission ready. Pre-combat checks are the responsibility of both individuals and leaders. Since it is impossible to overemphasize their importance, leaders should allocate a minimum of two hours in which to conduct them. Individual Marines are responsible for ensuring through pre-combat checks that all equipment necessary for the mission is on hand and serviceable. Pre-combat checks should be conducted by individuals, buddy teams, specialized teams, attachments, and squad leaders. Table A-1, table A-2, and table A-3 give examples of pre-combat checks checklists that may be utilized for individuals, crew-served weapons, and vehicles as part of squad leaders' mission preparations.

Table A-1. Pre-Combat Check Checklist (Individual).

	Medical/Hygiene Injuries Medical condition		Flashlight: • Serviceable • Red lens • Extra bulb(s)
	Fighting load Uniform (ID tags, card) Wrist watch Eye protection (ballistic) • Clear • Shaded Kevlar helmet (serviceable) Body armor • Serviceable • Correct protection level Fighting load carrier # magazines First aid kit complete Canteens Camelback Note-taking material (pen/pencil) All required gear "dummy corded"		Mission specific equipment: • Gas mask • Flex cuffs • Marking materials • Blackout goggles • Digital camera • Mechanical breaching equipment ♦ Wire cutters ♦ Sledgehammer ♦ Bolt cutters ♦ Hooligan bar • Extra batteries • Assault load • Sustainment load
	Ammunition: • Designated load-out (Individual) • Special ammunition spread-loaded		Weapons: • Clean, lubricated, and serviceable • Function checked • Optics • Zeroed • Serviceable • Sling • SL-3 complete
	Night vision devices: • Serviceable • Op-checked • Mounted properly (if applicable)		Communications equipment: • Serviceable • Correct encryption data • Extra batteries • Smart pack with call signs and frequencies • Spare antennas (including expedient)

Table A-2. Pre-Combat Check Checklist (Vehicle).

	Exterior body damage		Pioneer kit
	Serviceability: • Windshield • Windows • Mirrors • Wipers clean, functioning		Breaching kit
	Pintle secured		GPS with extra batteries
	Wheels/Tires: • Serviceable • Properly inflated • Firmly secured • Brakes function		Air panel
	Battery functioning		2 full fuel cans (stored externally)
	Engine:		2 full water cans
	• Transmission fluid level		Weapon cleaner/lubricant and kits
	• Oil level		1 case MREs
	• Power steering fluid level		Strip map with mission details for each driver
	• Coolant level		Spare petroleum, oils, and lubricants
			Power amp connected properly and functions
			Force continuum equipment in turrets: • Pen flares • Flash bangs • Pop-ups • Smoke grenades • Chem lights
	Brake fluid level		Thermite grenades (when available)
	Belts secured, serviceable		
	Lights: • Head & tail lights function • Infrared/blackout lights function		
	Tow bar Tow straps		Vehicle marked in accordance with operational requirements No extraneous markings or decorations
	Fire extinguisher		Personal entertainment removed (e.g. magazines, music devices)
	Spare tire (with jack and iron)		All equipment strapped down and stowed properly

Table A-3. Pre-Combat Check Checklist (Crew-Served Weapons).

Machine Guns			
	Function check, serviceability confirmed		Tripod serviceable
	Head space and timing confirmed		SL-3 complete • Cleaning gear • Spare barrel • Pintle
	T&E set on designated setting(s)		Optic mounted and zeroed
Mortars			
	Function check, serviceability confirmed		One compass per gun, compass orientation confirmed with other bearing device
	M64 has been bore-sighted		SL-3 complete
	Bipods serviceable, function checked		Unit call signs reviewed, Smart Pack carried by FDC
	Round count reviewed mortar rounds broken out, spread-loaded for transport, and allotted per gun		
Assault			
	Function checked and serviceability confirmed		Demo kit complete Crimpers Electrical tape Priming adapters
	Iron sight serviceable Function checked (zeroed) Range set		Required demolitions Broken out Inspected Prepared for transport Proper number of initiators Back-up initiators allotted and carried
	Scope bore-sighted		Rockets broken out, inspected, and spread-loaded

Pre-Combat Inspections

The responsibility for thorough and detailed inspections rests with small unit leaders. The responsibility to ensure that equipment is serviceable and that Marines possess the necessary knowledge to successfully accomplish their assigned missions cannot be delegated. Squad leaders' pre-combat inspections should spot check equipment readiness, but focus must be on individual Marines' understanding of the mission. Pre-combat inspections should validate that pre-combat checks have been performed. Leaders must make use of all available time to ensure pre-combat inspections are performed and that time is allotted for corrective actions to be performed. Thus, leaders must be competent and familiar in the maintenance and care of all unit equipment. The standards the leader sets determine the squad's ability to successfully complete assigned missions. Table A-4 is an example of a pre-combat inspection checklist.

Inspections are squad leaders' greatest assets to mitigate the enemies of unit readiness—boredom and complacency. It is human nature for Marines to get accustomed to their daily routines and begin to overlook minor issues. By ordering and conducting regular inspections, leaders are able to correct small problems before they become big problems.

Table A-4. Pre-Combat Inspection Checklist.

	What is the mission?		Show me the primary and alternate routes on your map or strip map.
	What is the commander's intent?		What are your tasks on this mission?
	What are the rally points?		What is the frequency and call sign for: Our unit The next higher unit Who we report to
	What are your collateral duties?		What is the challenge and password?
	What are the intelligence requirements for this mission?		What are the indirect fire targets along our route?
	How many personnel are on this mission?		What are the indicators of an IED or mine along the route?
	What are the rules of engagement?		How do we call the QRF?
	What has the recent enemy activity in our area been, including location and time?		What is the lost Marine plan?

REHEARSALS

Like pre-combat checks and inspections, rehearsals are important in preparing for assigned activities or tasks. Rehearsals do not check equipment, but rather individual Marines' knowledge and unit readiness. Rehearsals help ensure all members of the squad understand their assigned tasks and that they are capable of successfully performing their duties. Rehearsals provide practice prior to a mission; they also allow individuals and special teams to understand their specific assignments in support of the overall mission. Rehearsals are an important element of mission preparation that should always be completed prior to departing friendly lines. Table A-5 provides an example of a mission rehearsal checklist that squad leaders may utilize when conducting a squad-level rehearsal.

Table A-5. Mission Rehearsal Checklist.

Mission Planning Check/Rehearsals			
	Mission statement and commander's intent		Immediate action drills rehearsed: • Enemy contact/direct fire with enemy • Indirect fire • Obstacles • Aerial attack • CBRN attack • Electronic attack • Sniper fire • IED explosion/discovery • Near/far ambush • Crowd control • Foreign/American media
	Timeline reviewed		
	Collateral duties assigned		
	Info from leader's reconnaissance and S-2 incorporated into rehearsals		
	Routes reviewed (danger areas, rally points, checkpoints) • Primary • Alternate		
	Map overlay completed		CASEVAC rehearsal conducted
	Frequencies loaded; location of Smart Pack known by all Marines		Medical equipment checked: • IV fluid bags not expired • Squad's medical bag complete • Litters functional • Rigid • Pole-less • CASEVAC 9-line report format on every Marine
	Radio checks conducted: • Intra-squad • Higher • Supporting • Adjacent		
	Crypto equipment has current fill data and batteries for next changeover		
	Signal plan rehearsed		
	Rules of engagement reviewed		Personnel roster and EDL submitted to HHQ
	Bump plan reviewed (when vehicles apply)		Lost Marine plan reviewed and discussed in relation to rally points
LEGEND: EDL equipment density list IV intravenous			

APPENDIX B
HAND-AND-ARM SIGNALS

The following pages give explanations and diagrams of standard hand-and-arm signals utilized within Marine Corps tactical formations. This appendix does not cover all hand-and-arm signals, only those commonly used by ground combat element units. It is a guide only.

DECREASE SPEED
Extend the arm horizontally sideward, palm to the front, and wave arm downward several time, keeping the arm straight. Arm does no move above horizontal.

CHANGE DIRECTION OR COLUMN (Right or Left)
Raise the hand that is on the side toward the new direction across the body, palm facing front. Then swing the arm in a horizontal arc, extending arm and hand to point in the new direction.

ENEMY IN SIGHT

Hold the rifle horizontally, with the stock in the shoulder, the muzzle pointing in the direction of the enemy. Aim in on the enemy target and be ready to engage him/her if he/she detects your presence.

RANGE

Extend the arm fully toward the leader or people for whom the signal is intended with fist closed. Open the fist, exposing one finger for every 100 meters of range.

COMMENCE FIRING

Extend the arm in front of the body, hip high, palm down, and move it through a wide horizontal arc several times.

FIRE FASTER
Execute RAPIDLY the signal COMMENCE FIRING. For machine guns, this signals a change to the next higher rate of fire is required.

FIRE SLOWER
Execute SLOWLY the signal COMMENCE FIRING. For machine guns, this signals a change to the next lower rate of fire is required.

CEASE FIRING
Raise the hand in front of the forehead, palm to the front, and swing the hand and forearm up and down several times in front of the face.

ASSEMBLE

Raise arm and vertically to the full extent,
join fingers, palm to the front;
wave in large, horizontal circles.

FORM COLUMN

Raise either arm to the vertical position. Drop the arm to the rear, describing complete circles in a vertical plane parallel to the body. The signal may be used to indicate either a troop or a vehicular column.

ARE YOU READY?

Raise a hand in front of the forehead, palm to the front, and swing the hand and forearm up and down several times in front of the face.

I AM READY

Execute (echo) the signal for ARE YOU READY?

ATTENTION

Extend the arm sideways, slightly above horizontal, palm to the front; wave toward and over the head several times.

SHIFT

Point to individuals or units concerned; beat on chest simultaneously with both fists; then point to location you desire them to move to.

ECHELON RIGHT (Left)

The leader gives this signal either facing towards or away from the unit. Extend arms at 45 degrees above and below the horizontal, palms to the front. The lower arm indicates the direction of echelon. (Example: for echelon right, if the leader is facing in the direction of forward movement, the right arm is lowered; if the leader is facing the unit, the left is lowered.) Supplementary commands may be given to ensure prompt and proper execution.

SKIRMISHERS (Fire Team), LINE FORMATION (Squad)

Raise arms lateral and horizontal, palms down. To indicate a direction, move in that direction at the same time. When signaling for fire team skirmishers, indicate skirmishers right or left by moving the right/left hand up and down, regardless of the direction the signaler is facing. Skirmishers left will always be indicated by moving the left hand up and down; Skirmishers right, with the right hand.

WEDGE

Extend both arms downward and to the side at a 45-degree angle below horizontal, palms to the front.

VEE
Extend arms at a 45-degree angle above the horizontal, forming the letter V with arms and torso.

FIRE TEAM
Place the right arm diagonally across the chest.

SQUAD
Extend the hand and arm toward the squad leader, palm down; distinctly move the hand up and down several times from the wrist, holding the arm steady.

PLATOON
Extend both arms forward, palms toward the leader(s) or unit(s) for whom the signal is intended and make large vertical circles with hands.

CLOSE UP
Start signal with both arms extended sideward, bringing palms forward and together in front of the body momentarily. When repetition of this signal is necessary, the arms are returned to the starting position by movement along the front of the body.

OPEN UP, EXTEND
Start signal with arms extended in front of the body, palms together, and bring arms horizontal at the sides, palms forward. When repetition of the signal is necessary, the arms are returned along the front of the body to the starting position and the signal is repeated until understood.

DISPERSE
Extend either arm vertically overhead; wave the hand and arm to the front, left, right, and rear, the palm toward the direction of each movement.

LEADERS JOIN ME
Extend arm toward the leaders and beckon with finger as shown.

I DO NOT UNDERSTAND
Face toward source of signal; raise both arms sidewards, horizontal at hip level; bend both arms at elbows, palms up, and shrug shoulders in the universal "I don't know" manner.

FORWARD, ADVANCE, TO THE RIGHT (Left), TO THE REAR (used when starting from a halt)
Face and move in the desired direction of march; at the same time, extend the arm to the rear, then swing it overhead and forward in the direction of movement until it is horizontal, palm down.

HALT
Carry the hand to the shoulder, palm to the front; then thrust the hand upward, vertically extending arm fully, hold in that position until the signal is understood.

FREEZE
Make the signal for HALT, and form a fist.

DISMOUNT, DOWN, TAKE COVER
Extend arm sideward at a 45-degree angle above horizontal, palm down, and lower it to the side. Both arms may be used in giving this signal. Repeat until understood.

MOUNT
With the hand extended downward at the side, palm out, raise arm to a 45-degree angle above horizontal. Repeat until understood.

DISREGARD PREVIOUS COMMAND; AS YOU WERE
Face the unit or individual being signaled, then raise both arms and cross them over the head, palms to the front.

RIGHT FLANK
LEFT FLANK
(Vehicles, Craft, or Individuals Turn Simultaneously)
Extend both arms in direction of desired movement.

INCREASE SPEED; DOUBLE TIME
Carry the hand to the shoulder, fist closed; rapidly thrust the fist upward vertically, extending fully then back to the shoulder several times. This signal also used to increase gait or speed.

HASTY AMBUSH RIGHT
HASTY AMBUSH LEFT
Raise fist to shoulder level and thrust it several times in the desired direction.

RALLY POINT
Touch the belt buckle with one hand, then point to the ground.

OBJECTIVE RALLY POINT
Touch the belt buckle with one hand, point to the ground, and make a circular motion with the hand.

TO PREPARE FOR HELICOPTER GUIDANCE
Extend arms above the head, palms facing inboard.

TO DIRECT HELICOPTER FORWARD

Extend the arms and hands above the head, palms facing away from the helicopter. Move the hands in such a motion as to simulate a pulling motion.

TO DIRECT HELICOPTER TO EITHER SIDE

Extend one arm horizontally sideways in direction of movement and swing other arm over the head in the same direction.

TO DIRECT HELICOPTER TO LAND

Cross and extend arms downward in front of the body.

TO DIRECT HELICOPTER TO TAKE OFF
Make a circular motion of the right hand overhead, ending in a throwing motion toward the direction of takeoff.

TO DIRECT HELICOPTER BACKWARD
Extend the arms and hands, palms to the waist, facing the helicopter.
Move the hand to simulate a pushing motion.

TO DIRECT HELICOPTER TO HOVER
Extend arms horizontally sideways, palms downward.

TO DIRECT HELICOPTER TO WAVE OFF
Wave arms rapidly and cross them over the head.

TO DIRECT HELICOPTER TO HOOK UP EXTERNAL LOAD
Place the fists in front of body, left fist over the right, in a rope-climbing action.

TO DIRECT HELICOPTER TO RELEASE EXTERNAL LOAD
Left arm extended forward horizontally, fist clenched, right hand making horizontal sliding motion below the left fist, palm downward.

Appendix C
Orders Formats

Warning Order

A. Situation: ____________________

B. MISSION: ____________________

C. GENERAL INSTRUCTIONS:

NAME	CHAIN OF CMND	ELEMENT	SPECIAL TEAM	IND DUTIES	SPECIAL EQUIPMENT	GEAR COMMON TO ALL	TIME SCHEDULE			
						UTILITIES	WHEN	WHERE	WHAT	WHO
						BALLISTIC HELMET				
						BODY ARMOR				
						FLIC (COMPLETE)				
						BOOTS				
						GLOVES				
						BALLISTIC EYE WEAR				
						ID TAGS				
						ID CARDS				
						CAMELBACK				
						FIRST AID POUCH				
						PERSONAL WEAPON				
						WEAPON CLEANING GEAR				
						NVD(S)				
						CAMMO PAINT				
						MAP GEAR				
						COMPASS				

D. COORDINATING / SPECIFIC INSTRUCTIONS: ____________________

____

____

Figure C-1. Example Warning Order Format.

WARNING ORDER
SITUATION: ENEMY: (SALUTE/DRAW-D)
S-
A-
L-
U-
T-
E-
FRIENDLY: (HAS)
H-
A-
S-
ATTACHMENTS:
DETACHMENTS:
MISSION: (W.W.W.W.W.)
WHO:
WHAT:
WHEN:
WHERE:

WHY:

UNIT	Gen. Org.	Duties	Arms, Ammo. & Equipment	Gear Common To All	TIME SCHEDULE			
					WHEN	WHAT	WHERE	WHO

Specific Instructions:

Figure C-2. Example Warning Order Format—Continued.

Operation Order Format

REFERENCES: List any maps or documents needed to understand the order or that were used in the preparation of the order.

TIME ZONE USED THROUGHTOUT THE ORDER:

TASK ORGANIZATION:

1. SITUATION
   a. Enemy Forces
      1) Situation (size, activity, location, unit, time, and equipment)
      2) Capabilities (defend, reinforce, attack, withdrawal, delay)
      3) Most likely and most dangerous courses of action
   b. Friendly Forces
      1) Mission of your parent unit
      2) Mission of unit providing your support
      3) Mission and/or route of adjacent unit that may affect your mission
   c. Attachments and detachments

2. MISSION
   a. Who
   b. What
   c. Where (coordinates)
   d. Why

3. EXECUTION
   a. Concept of operation: The overall plan (scheme of maneuver) for the unit and plan for fire support.
   b. Commander's intent: How commander views the upcoming mission
   c. Sub-unit missions: for sections, teams, and individuals
   d. Coordinating instructions:
      1) Time schedule
      2) Formations and order of movement
      3) Route (primary and alternate)
      4) Movement within friendly front lines
      5) Rally points and actions at rally points
      6) Actions on enemy contact, at danger areas, and at the objective
      7) CBRN safety instruction and MOPP level and PIR requirements
      8) Fire support (if not previously discussed)
      9) Rehearsals and inspections
      10) Debriefing (including essential elements of information, and other intelligence requirements, time, and place).

Figure C-3. Operation Order Format.

Operation Order Format (Cont.)

4. ADMINISTRATION and LOGISTICS
    a. Rations / Chow plan
    b. Uniform, arms and ammunition
    c. First aid / medical plan
    d. Detainee / captured material handing plan

5. COMMAND and SIGNAL
    a. Command
        1) Squad leader's location
        2) Chain of command
    b. Signal
        1) Frequencies and call signs
        2) Pyrotechnics and signals
        3) Challenges and passwords
        4) Brevity and code words

Notes:
1. Details under subparagraphs should be tailored to provide all relevant and essential information.
2. Items covered by SOP need not be covered in the operation order, but should be referenced.

Figure C-3. Operation Order Format—Continued.

(Classification)

(Change from oral orders, if any)

Copy _ of _ copies
Issuing Headquarters
Place of issue (may be in code)
Date-time Group of signature
Message Reference Number

FRAGO

References:

1. SITUATION
   a. Enemy forces.
   b. Friendly forces.
   c. Attachments and detachments.

2. MISSION

3. EXECUTION
   Intent:
   a.Concept of operations.
   b.Tasks to maneuver units.
   c.Tasks to CS units.
   d.Coordinating instructions.

4. ADMINISTRATION AND LOGISTICS

5. COMMAND AND SIGNAL

(Classification)

ACKNOWLEDGE:

OFFICIAL:

NAME (Commander's last name)
RANK (Commander's rank)

ANNEXES:
DISTRIBUTION:

(Classification)

Figure C-4. Fragmentary Order Format.

APPENDIX D
CALL FOR FIRE

OVERVIEW

The MAGTF aggressively uses fire support assets to enable maneuver and create battlefield effects. The task of initiating and controlling fire support assets is assigned to a qualified JFO or JTAC when attached. However, all members of the rifle squad must become familiar with the protocols of fire support coordination so they can employ these powerful assets should the need arise. Basic fire support is controlled through formats standardized across the US military and NATO allies. These should be memorized and drilled as part of routine sustainment training.

SURFACE FIRES

The call for fire transmission is the basic command sequence the requesting agency transmits over the appropriate communications network to the firing agency. It is given in six basic elements: observer identification, warning order, target location, target description, method of engagement, and method of fire and control. These are sent in three distinct transmissions as shown in table D-1.

Table D-1. Call for Fire Sequence.

1st Transmission
Observer identification (call sign)
Warning order
2nd Transmission
Target location
3rd Transmission
Target description
Method of engagement
Method of fire and control

ELEMENTS OF THE CALL FOR FIRE

Observer Identification

This first element of the call for fire lets the fire direction center (FDC) know who is calling for fire and clears the net for the fire missions. The observer uses a call sign.

Warning Order

The warning order clears the net for the fire mission and tells the FDC the type of mission and the type target location that will be used. The warning order consists of the type of mission and the method of target location. It is a request for fire, unless prior authority has been given to order fire.

The type of mission can be any of the following:

- *Adjust fire*. When the observer believes the situation requires an adjusting round (because of a questionable target location), the observer announces, *adjust fire*.
- *Fire for effect*. When the observer is certain the target location is accurate and the first volley should have the desired effect on the target, so that little or no adjustment is required, the observer announces, *fire for effect*.
- *Suppress*. Suppression missions are typically fired on pre-planned targets, and the duration is associated with the call for fire.
- *Immediate suppression or immediate smoke*. When engaging a target that has taken friendly elements under fire, the observer announces, *immediate suppression* or, *immediate smoke*, followed by the target location.
- *Suppression of enemy air defenses*. Suppression of enemy air defenses (SEAD) fires neutralize, destroy, or temporarily degrade surfaced-based enemy air defenses by destructive or disruptive means.

Target Location

The observer provides the target location data to the FDC using the grid, polar, or shift from a known point method. The most common method uses grid coordinates. Grids are typically sent with a precision of at least six-digit grid coordinates. Eight-digit grids or higher precision should be sent if the observer's equipment can achieve a low target location error.

Target Description

This element contains sufficient detail (i.e., type, size, activity, and degree of protection) to enable the FDC to determine the amount and type of ammunition to use.

Method of Engagement

Observers use this element to describe how they desire to attack the target. The standard methods are area fire, low angle, high explosive/quick fuse, or circular sheaf.

Types of adjustment (area or precision fire) include—

- Danger close.
- Mark.
- Trajectory (low or high angle).
- Ammunition (projectile, fuse, or volume of fire).
- Distribution (converge, open, linear, rectangular, or irregular sheaf).

Method of Fire and Control

In this element, the observer indicates the desired manner of attack, the method of fire, and who has fire control authority. The observer also indicates the ability to directly observe the target. Options include the following:

- *Fire when ready*. This is the default unless otherwise specified. Firing commences when the firing unit is ready.
- *At my command*. The firing agency will indicate when it is ready but will not commence firing until the observer gives the fire command.
- *Continuous illumination*. This instructs the firing agency to continue firing illumination rounds until told otherwise.
- *Continuous fire*. The firing agency will fire until told otherwise. This command can also be accompanied with a desired rate of fire (e.g. one round every 30 seconds).
- *Repeat*. This instructs the firing agency to repeat the previous fire mission.

> ***Note:*** This is why the word "repeat" is never used in the course of normal radio traffic.

- *Request splash*. This requests that the firing agency broadcast, *splash!* approximately five seconds prior to the round's impact.
- *Duration*. The firing agency will fire continuously for the specified duration.

Message-to-Observer

The message-to-observer is information sent to the observer by the supporting FDC. The observer acknowledges the message-to-observer by reading it back in its entirety. If conducting a grid or shift mission, the observer must provide the direction from themselves to the target after the read-back (e.g., *break, 1400 mils*).

Message-to-observer information includes—

- Units to fire (firing unit, adjusting unit)
- Changes to call for fire (if any)
- Number of rounds (per tube)
- Target number
- Time of flight (in seconds)
- Ordinate altitude (highest point of the rounds flight) information

After spotting the weapons' impacts and effects, the observer should send corrections to the FDC to move the burst onto an adjusting point. The observer sends corrections in meters in the reverse order of that used in spotting (i.e., deviation, range, and height of burst). The following is an example of the elements contained in the adjustment transmission.

Example: Elements Contained in the Adjustment Transmission

Adjustments:
Left, or *Right* (meters from impact to observer-target line [OTL])
Add, or *Drop* (meters, distance from impact to target)
Up, or *Down* (meters, distance from height of burst to desired height of burst)
Fire for effect, over. (Sent with final correction when desired effects on target are observed).

After the fire mission is complete, the call for fire concludes with a mission complete transmission. The following provides examples of end-of-mission transmissions.

Example: Mission Completion Transmissions

End of mission, (battle damage assessment and target activity reported to FDC), *over.*
-or-
Refinements (if any),
Record as target (for later use or reference),
End of mission (see above),
Surveillance (indicates observer will continue to watch the target area)

OTHER CALL FOR FIRE TERMS AND DEFINITIONS

Danger Close

This is included with the method of engagement when the predicted impact of a round or shell is within 600m of friendly troops for mortars and artillery, and 750m for naval surface fires. The creeping method of adjustment (i.e., no adjustment greater than 100m) is used exclusively during danger close missions. The "danger close" method of engagement should not be confused with risk estimate distances or minimum safe distances.

Dark Star

An illumination round that fails to ignite or deploy a parachute properly.

Direction

The direction from the observer to the target. It is usually transmitted in mils grid; degrees may be transmitted, but must be announced so the FDC is aware that degrees are being used vice mils.

Fresh Target

This is an order that can be sent any time during a fire mission to indicate that a spotter needs to engage a higher priority target. The call for fire begins with, *fresh target*, and the ship interrupts fire on the original target to engage the fresh target. The target location is sent as a correction from the last impacted salvo of the original target (i.e., shift from a known point) and must include any elements of the call for fire which differ from the original target.

Mark

This orders the firing agency to fire a spotting round, typically white phosphorous or illumination, and is used to visibly indicate targets to aircraft, ground troops, or fire support.

Maximum Ordinate (Maxord)

In artillery and naval gunfire support, this is the height of the highest point in the trajectory of a projectile above the horizontal plane passing through its origin. When announced, personnel should be sure to correctly describe the unit of measure (e.g., meters or feet) used. Conversion may be required.

New Target

An order that can be sent any time during a fire mission to indicate that the spotter requires the firing agency to engage a target not necessarily of a higher priority than the one already being engaged. The call for fire begins with, *new target*, and the ship continues to fire on the original target. The target location can be sent using any of the standard methods, and any elements of the new target which differ from the original target must be sent.

Salvo

This is one shot fired at a target simultaneously by all or part of the guns in a battery.

Observer to Target (OT) Factor

This is the distance in meters from the observer to the target, expressed to the nearest thousand meters and in thousands. When the distance is greater than 1,000m, the observer to target distance is determined to the nearest 1,000m. If the distance is less than 1,000m, the distance is determined to the nearest 100m and expressed in thousands. For example, at a distance of 2,322m, the observer to target factor would be expressed as a 2; for a distance of 800m, it would be expressed as .8.

Ordinate (ORD) "X"

This is defined as the altitude of an artillery round at a specific distance from the target along the GTL back toward the firing location. For example, the altitude at two kilometers toward the tubes from the target would be expressed as *ORD-2*.

Repeat

While adjusting fire, this means for the firing agency to fire one round again using the same method of fire. During a fire for effect, it means for the firing agency to fire the same number of rounds using the same method of fire.

Ripped Chute

This is sent by the spotter to indicate that an illumination round parachute was ripped or separated on deployment.

Rounds Complete
This indicates that the fire for effect stage is complete.

Shot
This is communicated by the FDC to indicate that one or more rounds have been fired.

Splash
This is passed by the FDC about five seconds prior to the estimated time of impact to inform the observer or spotter.

Time on Target
The time on target (TOT) indicates the time the observer desires one or more rounds to impact.

MISSION FORMATS

The mission formats in the following examples demonstrate call for fire transmissions when conducting grid, polar, or shift from a known point missions.

> ***Note:*** An accurate position report is needed by the firing agency prior to the call for fire transmission in order to plot targets and make the correct adjustments. While grid missions can be executed without an updated position report, it is necessary before any adjustments can be made.

Example: Adjust Fire Mission (Grid)

(FDC call sign) *this is* (observer call sign), *adjust fire, over.*
Grid ____________ (min. six digits), *over.*
Target Description: (type, size, activity)
Method of engagement:____________
Method of fire and control:_________
Over.

Example: Adjust Fire Mission (Polar)

(FDC call sign) *this is* (observer call sign), *adjust fire polar, over.*
Direction (OTL to the nearest 10 mils or 1 degree).
Distance (to nearest 100m).
Up/Down (to nearest 5m).
Target Description: (type, size, activity)
*Method of engagement:*____________
*Method of fire and control:*_________
Over.

Example: Adjust Fire Mission (Shift from Known Point)

(FDC call sign) *this is* (observer call sign), *adjust fire shift* (known point identification) *over.*
Direction (OTL to the nearest 10 mils or 1 degree).
Left /Right (lateral shift to nearest 10m).
Add /Drop (range shift to nearest 100m).
Up/Down (to nearest 5m).
Target Description: (type, size, activity)
*Method of engagement:*___________
*Method of fire and control:*_________
Over.

ADJUSTING FIRES

To conserve ammunition and ensure the accuracy of fires, adjust fire missions are used to ensure fires are able to create the effects desired. After calling for an adjust fire mission, the firing agency fires one salvo of rounds. The observer checks their impacts and determine whether or not the sheaf is accurately placed to generate the desired effects on target.

> ***Note:*** The sheaf of round impacts represents a probable grouping of fire-for-effect impacts. Therefore, direct hits on target are not necessarily required.

As most fires have explosive area effects, all that matters is that the sheaf placement maximizes the probability that effects on target will be generated. Therefore, to properly estimate their effects, one must understand the effective casualty radius and other characteristics of the munitions being used.

If adjustments need to be made, the observer reports them to the firing agency as shown in the following example.

Example: Adjust Fire Mission (Corrections)

(FDC call sign) *this is* (observer call sign),
Left /Right ________,
Add /Drop ________,
Up/Down _________.
Over.
Once effects on the target are created:
(FDC call sign) *this is* (observer call sign). *Fire for effect, over.*

Observer-to-Target Factors

When making adjustments, it is important to recognize the importance of the distance from the observer to the target. When using binoculars with a range estimation capable reticle, a compass, or other spotting tools, the increments (usually mils) on the instrument are very helpful in relaying accurate adjustments to the firing agency. However, the distance to the target can magnify or diminish the readings on spotting instruments. For this reason, it is important for the observer to account for the observer-to-target factor, commonly referred to as the OT factor. This is a simple technique used in adjusting bursts based on the distance from the observer to the target. The observer-to-target factor is rounded up to the nearest 1,000m. If the range to the target is less than 1,000m, the observer-to-target factor is a decimal reflecting the nearest 100m. Examples of observer-to-target factors are shown in table D-2.

Table D-2. Observer-to-Target Factor Calculations.

Distance (meters)	OT Factor
500	0.5
1000	1
1700	2
2000	2
3000	3

Example: The center of the observed sheaf on the initial volley of an adjust fire mission lands 100 mils to the left of the target when observed through binoculars. The range to the target is 2,000m. The proper adjustment to be given to the firing agency is: *right 200.*

Note: All adjustments are to be rounded to the nearest 10m.

The formats in the following examples are utilized when conducting a fire for effect, suppression/ obscuration, or marking mission using the grid method.

Example: Fire for Effect Mission (Grid)

(FDC call sign) *this is* (observer call sign). *Fire for effect, over.*
Grid (min. six digits).
Target Description: (type, size, activity)
*Method of engagement:*____________
*Method of fire and control:*__________
Over.

Example: Suppression, Obscuration Missions (Grid Method)

(FDC call sign) *this is* (observer call sign). *Suppress/immediate suppression/immediate smoke* ________(target number or min. six digit grid), *over.*

Note: Observer may include a "DURATION" call after target location to specify time or suppression or obscuration.

Example: Marking Mission (Grid Method)

(FDC call sign) *this is* (observer call sign). *Fire for effect, over.*
Grid (min. six digits), *over.*
Marking round, white phosphorous, at my command, request time of flight, over.

Note: Munition type and method of control may vary.

Suppression of Enemy Air Defenses

Suppression of enemy air defenses missions is used to neutralize, destroy, or temporarily degrade enemy air defense capabilities while providing a marking round for aviation assets to locate their target. They instruct the firing agency to interrupt their firing for the TOT (i.e., the time during which the aircraft will deliver its munitions) of the aircraft being employed, as a safety measure to deconflict airspace.

The key difference in each mission format is in the method of engagement. The observer must dictate which timeline for surface fire impacts is required. The three timelines are—

- Continuous. Continuous fire directs the surface firing agency to create effects on target at TOT minus 60 seconds (i.e., TOT-60), TOT-30, TOT, TOT+30, and TOT+60.
- Interrupted. Interrupted fires only have impacts at TOT-60 and TOT-30.
- Nonstandard. Nonstandard means that a custom timeline is provided by the observer that is driven by the situation.

Note: The TOT must be confirmed with the aviation asset prior to transmission of the SEAD mission. The geometries and physics of flight and the desired end state of the ground tactical commander determine what is possible in terms of TOT. That timeline should then be used to drive the SEAD mission.

The format for SEAD (pronounced See-Ad) missions is as follows:

Example: Suppression of Enemy Air Defenses (SEAD) Format

(FDC call sign) *this is* (observer call sign). *SEAD, over.*
Grid to suppress: (6-digit min.), *grid to mark* (6-digit min.), *over.*
Target Description: (continuous/interrupted/nonstandard, CAS TOT_______), *over.*

AVIATION-DELIVERED FIRES

Close air support (CAS) involves employing fire support from rotary-wing or fixed-wing aircraft. Because of the technical nature of aviation and the various deconfliction it requires, a dedicated controller is often assigned to manage these assets. The squad leader typically works with a forward air controller, JTAC, or JFO to employ CAS. However, it is crucial that all Marines in the squad understand the basics of CAS employment.

Pilots are trained to work with unqualified observers and will work with Marines to provide effective and safely employed ordnance in support of the mission. At a minimum, all Marines should be familiar with the following elements of the CAS execution checklist in table D-3.

Table D-3. CAS Execution Checklist.

CAS aircraft check-in
Situation update
Game plan
CAS brief
Remarks/restrictions
Read-backs
Attack
Battle damage assessment (BDA)

Close Air Support Aircraft Check-in

When a friendly aircraft checks in to the battlespace or is handed off to an observer/controller, the aircraft will provide the controller a brief on its capabilities and limitations.

The general format for a CAS aircraft check-in is as follows:

CAS Aircraft Check-in Format

(Controller call sign) *this is* (aircraft call sign). *Mission number* _____
Number and type of aircraft
Position and altitude
Ordnance (type, fuzing)
Time on station
Abort code
Remarks

Situation Update

This brief is given from the controller to the aircraft. It is tailored to the tactical situation and covers the amount of time the aircraft is expected to be on station. This plain-language brief should cover the following:

- Threats: Specifically include any surface-to-air threats.
- Enemy situation.
- Friendly situation/commander's intent.
- Indirect fire activity: Include friendly firing agency locations, frequencies, call signs, GTLs, and max ordnances.
- Clearance authority: When no forward air controller, JTAC, or JFO is present, control may be delegated to the senior ground tactical commander. In this case, the controller must inform the aircraft that they are not a JTAC.
- Hazards: Weather, terrain, or other obstructions to flight must be passed.
- Remarks/restrictions.

Game Plan

The controller gives the game plan brief to the aircraft. This plain-language, succinct plan of action, covers the overall scheme of maneuver and dictates the concept of coming events. The controller avoids repeating information given in other portions of the control, and includes the—

- Type of control (i.e., type I, II, or III as shown in table D-4).
- Method of attack (i.e., bomb on target or bomb on coordinate).
- Effects and/or ordnance desired.

Table D-4. Control Type Descriptions.

Type of Control	Description
Type I	Controller can observe both the target and the aircraft.
Type II	Controller can observe either the target or the aircraft.
Type III	Controller cannot directly observe the target or the aircraft.

Close Air Support Brief

This portion of the brief includes the NATO standard CAS 9-line brief shown in table D-5.

Table D-5. NATO CAS 9-Line Brief Information.

Initial point (IP)/battle position (BP)
Heading in degrees magnetic from IP/BP to target
Distance. IP to target in nautical miles; BP to target in meters
Target elevation in feet (above mean seal level, or MSL)
Target description
Target location
Type of mark/saser code
Location of friendlies from target in cardinal direction and meters
Egress direction

Note: Initial points are used for fixed-wing CAS; battle positions are used for rotary-wing CAS.

Remarks/Restrictions

This portion of the brief is given to the aircraft by the controller and contains any other pertinent data not previously passed to the aircraft, such as—

- The desired type and amount of ordnance.
- The location and type of SEAD mission running concurrently.
- GTLs for surface fires in the area of operations.
- To designate a requested final attack heading (i.e., a window of at least 45 degrees).
- To request for a TOT.
- Requests for calls from the aircraft.

The following is an example of a brief for fixed-wing CAS:

:

Example: Fixed-Wing CAS Brief

India 1/1: *Hawg 11, India 1/1, advise when ready for game plan.*

CAS aircraft: *India 1/1, Hawg 11 ready for game plan.*

India 1/1: *Hawg 11, this will be a Type 2, BOT, 2x Mk-82 each aircraft, 30 second separation, advise when ready for 9-Line.*

CAS aircraft: *India 1/1, Hawg 11 ready for 9-Line.*

> ***Note:*** The 9-line brief is usually transmitted three lines at a time to allow the aircrew to copy the required information.

India 1-1: *MAZDA*
360
Nine point nine"...

*"Four five zero feet**
2xBTR-60s dug in
*NB 863 427"...**

"WP [White Phosphorous]
South 900
Egress east to CHEVY 17-19 thousand
Advise when ready for remarks and restrictions.

CAS aircraft: Ready to copy remarks and restrictions.

India 1/1: *Final attack heading 300-345, gun target line 280, TOT 32.**

CAS aircraft: *Four five zero feet, NB 863 427,*
final attack heading 300-345, TOT 32.

Note: Line 4, line 6 and restrictions (denoted with *) are mandatory read-backs. The controller may request additional read-backs.

Attack

Once the attack phase has commenced, communication priorities should go to the controller and the CAS aircraft. The aircraft will announce its actions in sequence and the controller will be responsible for issuing the commands listed in table D-6, as appropriate.

Table D-6. Controller Commands for CAS Aircraft.

Command	Action Taken
Abort	Abort the pass. Do not release ordnance. Abort calls are not restricted to the controller.
Cleared Hot	Aircraft is cleared to release ordnance on this pass.
Cleared to Engage	Type III only. Attack aircraft may initiate attacks within parameters specified by the controller. Aircraft will provide *commencing-engagement* and *engagement-complete* calls.
Continue	Aircraft is authorized to proceed with the attack profile but does not have clearance to release ordnance yet. This is used as an acknowledgment to the aircraft without providing clearance.
Continue Dry	Continue the present maneuver, ordnance release not authorized. This is used to approve the aircraft to continue its pass without expending ordnance.

Battle Damage Assessment

This call is given to the aircraft by the controller; at a minimum, it informs the aircraft whether the pass was successful or unsuccessful, or if the desired effects of the commander's intent were created. If possible, the type and number of targets destroyed, neutralized, and/or suppressed should be passed to the aircrew.

> *Note:* Rotary-wing assets may allow the controller to pass an abbreviated five-line CAS brief in place of the full CAS 9-line brief, as follows.

Rotary-Wing Abbreviated 5-Line CAS Aircraft Check-in Format

Observer/Warning Order/Game Plan
(Aircraft call sign) *this is* (controller call sign). *Five-line,*
Type I/II/III Control
MOA (BOT/BOC)
Ordnance requested.

Friendly Location/Mark
My position (grid/target reference point), *marked by* (strobe, panel, etc.)

Target Location
Target location (grid, degrees magnetic from controller and range in meters, or target reference point)

Target Description/Mark
Target (description), *marked by* (IR pointer, tracer, WP, etc.)

Remarks/restrictions (see 9-line)

APPENDIX E
SITE EXPLOITATION

Site exploitation consists of systematic actions aided by the appropriate equipment to ensure that personnel, documents, electronic data, and other materials at a site are identified, evaluated, collected, and protected in order to gather intelligence and facilitate follow-on actions.

Intelligence drives operations. But where does it come from? The answer is, every Marine is a collector. Making the transition from combat to investigation (similar to what crime scene detectives do) is a critical skill during any military operation. Once the fighting ends, evidence must be collected and preserved. The information gained from site exploitation is used in the intelligence cycle to drive follow-on operations and may be used by US, coalition, or host nation authorities to prosecute detainees. Methodical preparation and execution ensures success. Table E-1 provides a checklist of considerations and actions to take during site exploitation.

Table E-1. Site Exploitation Checklist.

	Systematic Search	
	Maintain site security and be cautious of layered security (i.e. booby traps).	Expedite items of high intelligence value to the rear for processing.
	Pre-designate teams, rehearse, and equip with: • Rubber gloves • Evidence bags/tags • Digital camera • Note taking material • Flashlights • Voice recorders • Explosive residue kits • Flex cuffs • Extra goggles - "blacked out".	Recover all material of potential intelligence value that can— • Identify follow-on targets • Paint localized intel picture • Confirm or negate intelligence requirements (IRs) and priority intelligence requirements (PIR)s • Provide evidence for conviction of detainees.
	Search method: • Circle • Grid • Zone/sector	Photograph/video everything prior to, during, and after searching.
	Identify individuals to be detained.Only trained personnel conduct tactical questioning, and in accordance with ROEs and S-2 guidance.	Photograph detainees with located contraband for judicial proceedings.
	Document the chain of custody for all detainees and evidence.	Sketch the floor plan noting where individuals and evidence were located.
	Record verbal and written statements from Marines and locals.	Conduct a thorough debrief with the CLIC as soon as possible after returning from mission.
When in doubt – Document!		

APPENDIX F
FIELDCRAFT

OVERVIEW

Fieldcraft consists of the tactical skills and methods necessary to survive and operate in daytime and periods of reduced visibility in all types of terrain. These skills include position selection, packing techniques, observation, shelter construction, and camouflage, among others. Good fieldcraft is essential for the effectiveness and survival of the Marine rifle squad. Efficient fieldcraft is only possible by application through continuous training in various conditions. In a tactical environment, fieldcraft is a 24-hour per day concern.

COVER AND CONCEALMENT

It is important to understand that cover and concealment are two different things. Just because Marines have concealment, does not mean they have cover. A bush or tall grass may provide excellent concealment but little to no cover from enemy fire.

Cover

Cover is protection from the fire of enemy weapons. It may be natural or man-made. Natural cover includes logs, trees, stumps, ravines, hollows, and reverse slopes, among others. Man-made cover includes fighting positions, trenches, walls, rubble, abandoned equipment, and craters. Even the smallest depression or fold in the ground gives some cover; this is known as micro-terrain. Marines must learn to look for and use every bit of cover the terrain offers.

Concealment

Concealment is anything that hides Marines and their position, unit, or equipment from enemy observation. Small unit leaders must enforce light and noise discipline, control movement, and supervise the use of camouflage. Well-hidden fighting positions help conceal the squad's location from the enemy.

The best way to use natural concealment is to refrain from disturbing the natural vegetation when moving into an area. Darkness alone does not conceal a unit from an enemy who has night vision or other detection devices.

CAMOUFLAGE

Camouflage is the use of concealment and disguise to minimize the possibility of detection and/or identification of troops, material, equipment, and installations. The purpose of camouflage is to conceal military objects from enemy observation. Camouflage is also used to conceal an object by making it look like something else. A rifle squad's mission often requires individual and equipment camouflage. Camouflage makes use of both natural and man-made material. Used well, it reduces the chance of detection by the enemy. When used properly, branches, bushes, leaves, and grass provide the best camouflage. Foliage used as camouflage must blend with that of the surrounding area. Individuals, gear, and exposed positions can be concealed from enemy observation by using the right materials and procedures.

When using camouflage, remember that objects are identified by their shape (i.e., outline), shadow, texture, and color. The principle purpose of camouflage in the field is to prevent direct observation and recognition. Some things the enemy will look for in trying to find friendly forces are described below.

Movement

Movement draws attention. An observer will catch movement in their field of view. Movement, such as hand-and-arm signals, can be seen by the naked eye at long ranges. Sudden movement attracts the eye. Slow and careful movement is much less likely to disclose the location of a well concealed position than quick and short movement.

Shadows

Shadows draw attention. Camouflage should be used to break up the shadows of fighting positions and equipment. Shaded areas offer concealment. There are two types of shadows: cast shadows and contained shadows.

Cast Shadow. In sunlight or moonlight, an object casts a shadow which may indicate its presence. An object which is concealed in other shadows is harder to detect since it does not cast a shadow of its own. As the sun or moon moves, so do the shadows. Objects which were concealed by shadow before may be revealed as the shadow moves. They may also be revealed by their own distinctive shadow, which reappears.

Contained Shadow. A contained shadow is contained within a space, for example, in a room, a cave mouth, or under an individual shelter. It is typically darker than other shadows and can therefore attract attention.

Shape

Shape is the outline of something. Distinct shapes are easily noticed. The shape of a helmet is easily recognized, as is the undisguised shape of a person's body. Camouflage must be used to conceal shapes which are distinctive and familiar to help them blend with their surroundings.

Shine

Shine may be a light source such as a cigarette glowing in the dark, or reflected light from smooth, polished surfaces such as a worn metal surface, a windshield, binoculars, eyeglasses, a watch crystal, or exposed skin not toned down with camouflaged paint. The use of lights or the reflection of light may help the enemy detect friendly positions. Equipment that shines should be subdued or covered with mud or paint.

Color

Contrasting colors are more easily detected. For example, light colors contrast against the dark green of jungle foliage, and a white skin complexion is more prominent than black. Camouflage should match the surrounding area rather than contrast with it. Bright colors should not be used in camouflage.

Dispersion

Dispersion is the distance between Marines, vehicles, or equipment. If a squad is not dispersed, it is easier to detect and hit. Distances between individuals, teams, and squads must be prescribed and enforced. As the terrain changes, the dispersion must be changed to match it.

When camouflaging, Marines should use only that material which is needed. Too much camouflage may call attention to a person or position as easily as too little will. Camouflage materials should be gathered from a wide area. An area stripped of all of its foliage will draw attention. When camouflaging, the use of black should be minimal or omitted completely. Very few objects in the natural environment are truly black in color.

Individual Camouflage

Successful individual camouflage involves the ability to recognize and take advantage of all forms of available natural and artificial concealment (e.g., vegetation, soil, debris). Marines must learn how to read the terrain and be able to determine how best to blend into their surroundings. Each Marine must use terrain to give themselves cover and concealment. They must supplement natural cover and concealment with camouflage. Before camouflaging, Marines study the terrain and vegetation of the area they are in and the area to which they are going. Grass, leaves, brush, and other natural materials must be arranged to blend with the surrounding area. Vegetation changes from area to area; as Marines move from one area to another, camouflage must be changed to blend with the surroundings.

Helmet

The helmet is camouflaged by breaking up its shape, smooth surface, and shadow. Helmets should be covered with the issued helmet cover, augmented by cloth, burlap, or camouflage netting colored to blend with the terrain. The cover should fit loosely to break up its natural shape. Foliage can be placed into the slits manufactured into the helmet cover and should stick out over the edges. When the Marine is moving, foliage should be draped, not plumed, as head movement will give away the Marine's position. This should not be overdone. Plumed foliage should only be added to the helmet if the Marine is performing a task that requires minimal movement (e.g., manning a fighting hole, LP/OP, ambush site). Bushes do not walk around; therefore, moving bushes are easily detected. If there is no material for helmet covers, the form and surface of helmets can be disguised and dulled with irregular patterns of paint or mud. Camouflage bands, parachute cord, burlap strips, or rubber bands can be used to hold foliage in place.

Clothing

Uniforms must blend with the terrain. If camouflaged clothing is not available, other available clothing can be attached in irregular splotches of appropriate colors. When operating in snow-covered terrain, Marines should wear over-whites. If over-whites are not issued, sheets or other white cloth can be used. When executing tasks that require minimal movement (e.g., ambush positions) a ghillie suit can be constructed if time, material, and the surroundings permit (see MCTP 3-01E, *Sniping*). Marines may improvise an expedient ghillie suit using burlap strips or foliage held in place by using boot bands, rubber bands, or other materials.

Body

Exposed skin reflects light and draws the enemy's attention. Even very dark skin reflects light because of its natural oil. The buddy system is recommended when applying camouflage face paint. Prior to camouflaging, Marines must study the terrain to determine which colors and patterns are best suited for the environment they are operating in. There are three techniques used to apply camouflage paint (see figure F-1):

- Splotching.
- Striping.
- A combination of splotching and striping.

Figure F-1. Camouflage Paint Application Techniques.

When applying camouflage paint, a combination of two or three colors should be applied in an irregular pattern to disguise areas of shine and shadow. Shine areas are more pronounced (e.g., forehead, cheekbones, nose, ears, and chin) and are painted with a dark color. Shadow areas are more recessed (e.g., around the eyes, under the nose, and under the chin) and are painted with a light color.

The end result should create a flattened appearance, which helps to conceal the distinct shapes of the face. All exposed skin should be painted, including the hands, back of the neck and head, ears, eyelids, and lips. If a Marine has light hair that is visible when wearing their helmet, camouflage paint should be applied to all exposed hair. Since the color black is rarely found in nature, the use of black paint should be avoided. However, black paint may be used to darken other colors to better match a Marine's surroundings.

It is important to select colors that best blend in to a Marine's surroundings, as determined during a study of the terrain. Colors best suited for various environments include—

- Woodland (i.e., green and brown).
- Desert (i.e., brown and tan).
- Snow-covered (i.e., white and any other colors that may match exposed vegetation and terrain).

If camouflage paint is not available, mud, charcoal, burnt cork, or other expedient means can be used to tone down exposed skin. Mud should be used only in an emergency because it changes color as it dries and may peel off, leaving the skin exposed. Since mud may contain harmful bacteria, mud should be washed off as soon as possible.

Individual camouflage is a continuous action. Perspiration and exposure to rain or wet elements can cause the paint to wear off over time. Leaders and Marines must be aware of these conditions and continuously check both themselves and each other to ensure they remain properly camouflaged, reapplying paint as necessary.

CAMOUFLAGING EQUIPMENT

Any equipment that reflects light should be covered with a non-reflective material that aids in its concealment. Badly worn or faded equipment may be hard to conceal. Units should turn in badly faded equipment or use mud, paint, or other materials to color it until it can be exchanged.

Weapons

The straight line of rifles or other infantry weapons may be very conspicuous to an enemy observer. The barrel and hand guard should be wrapped with strips of contrasting colored cloth or tape to break the regular outline. Camouflage paint, mud, or dirt dulls the reflective surface of the stock, barrel, and bayonet where coloring has been worn down. Burlap or foliage may be held in place to hand guards by use of boot bands, rubber bands, or similar materials. When camouflaging a weapon, its function must not be impaired.

Pouches, Packs, and Straps

Pouches, packs, and straps may be painted in irregular patterns to better match the terrain the Marine is operating in. When staging gear for a prolonged period (e.g. during an ambush or prior to an assault), natural vegetation may be inserted through straps and webbing to better conceal it among the terrain.

Buckles, Snaps, and Zippers

Metal reflects light. Metal fasteners such as buckles, snaps, and zippers should be subdued using paint or tape.

CAMOUFLAGING POSITIONS

Camouflaging individual Marines' positions and slit trenches is a continuous action that occurs prior to, during, and after entrenching.

Prior

Camouflaging a position begins before foliage is cleared or dirt is broken through proper site selection. When selecting a position, squad leaders must ensure that the proposed position will not be easily identified by passing enemy forces or civilians. Care must be taken to avoid silhouetting; if possible and avoid placing positions on the side of a hill or on a military crest.

During

Marines must work to camouflage their positions as they prepare them. Positions should blend into their surroundings. Excessive clearing should be avoided. When possible, open areas should be covered by fire rather than physically occupied, as it is harder to conceal a position in the open. Using too much material when camouflaging should be avoided, whether natural or man-made. Too much material can make the object and its shadow stand out from its surroundings, attracting the attention of a passing observer. Dirt must be camouflaged or hidden. If the position has no overhead, its bottom must be camouflaged to prevent detection by aircraft. Work on a position in daylight depends on the enemy air threat and whether or not the enemy can see the position. Marines should vary their routes to and from their positions to avoid creating a trail in the foliage that could give away their positions. During their construction, Marines should periodically move forward of their positions and look at them through the eyes of potential adversaries to ensure they will not be easily detected.

After

Camouflage must be constantly maintained after the position is completed. Over time, dead foliage wilts and changes color, which can easily attract the attention of enemy or civilians. Replacing wilted foliage should be done during periods of reduced visibility.

WATERPROOFING

Marines operate in every type of terrain and climate throughout the globe. They train, operate, and live outdoors. Bodies of water, periods of heavy rain, and high humidity are commonly encountered by the Marine rifle squad. The equipment Marines operate with is regularly carried on their bodies, and thus it must be properly protected and cared for. When operating in wet conditions, proper waterproofing is vital to the Marines' ability to execute their mission.

Clothing and Sleeping Systems

Wet clothing and sleeping systems can lead to a wide range of adverse effects on the individual Marine. Trench foot, jungle rot, and hypothermia are all conditions that commonly occur when a Marine is exposed to wet clothing for prolonged periods. Marines must work to keep themselves clean and dry to help prevent these conditions. Sleeping systems should be properly packed in the issued waterproof compression bags. If they become wet during use, they should be air dried at the soonest opportunity through use of the issued mesh bag or other means. Clothing should be packed in the issued waterproof compression bags. If compression bags are unavailable, sealable plastic bags or other expedient waterproof containers may be utilized.

Maps

Maps are vital to the rifle squad, as they are often the only reference Marines have for their location. Maps commonly depict routes, landmarks, and known or suspected enemy and friendly positions. To ensure their longevity, maps must be properly waterproofed. Laminating maps is ideal; this allows Marines to reuse the map from mission to mission by sanitizing any marks while keeping the map safe from water damage. If the means to laminate maps are unavailable, the Marine must improvise with the materials available. Maps may be covered in layers of clear tape or placed inside of a clear sealed bag while still retaining many of the benefits of lamination.

Radios

Radios are a vital part of the Marine rifleman's equipment. Radios are often the rifle squad's only source of communication with other friendly units, which it relies on for reinforcement, fire support, and CASEVAC. Thus, it is critical that the radio be protected. Military radios are designed to be water resistant; however, exposure to deep water (usually in excess of 10 feet) and normal wear during prolonged use can result in the diminishing of the radio's ability to resist water. When operating in wet conditions, radios should be wrapped in a water-resistant material, with any openings taped off with waterproof tape. Trash bags and duct tape are ideal for this type of expedient waterproofing.

Other Equipment

As resupply can be limited, it is vital that every piece of the rifleman's gear that may be detrimentally affected by wet conditions is protected. It is important that the gear's ability to absorb and retain unnecessary water be minimized to avoid needlessly adding to the weight of the Marine's load. Excess weight can slow the unit's speed of movement and contribute to the individual rifleman's exhaustion and fatigue. If a piece of gear can absorb and retain water, measures should be taken to prevent water from reaching it.

PACKING

The amount of thought and care Marines take in packing equipment can have significant effects on their ability to execute and accomplish their assigned mission. When packing, leaders should study the METT-T analysis and issue a warning order for the upcoming mission. Marines should pack only what has been deemed essential to successfully accomplish the mission. Special consideration should be given to gear that is most needed, and it should be packed in a way that allows the Marine to easily access it as required. Marines must be aware of where each piece of equipment is located (in accordance with unit SOP) so that they can access it as quickly and easily in the dark as they can during the day.

A well-loaded pack should feel stable and balanced when resting on the hips and prevent the contents from shifting inside. The heaviest items should be packed on top and close to the back to center the pack's weight over the hips and assist the Marine in maintaining an upright stance. For the purpose of packing, the pack can be divided into three sections—the bottom, the center or core, and the top and periphery (see figure F-2).

Bottom
Less frequently used and lighter-weight items should make up the bottom load of the rifleman's pack (e.g. sleeping system, extra utilities, and warming layers). The issued sleeping mat should be strapped to the bottom of the pack.

Center
Heavier and more frequently used items should be placed in this portion of the pack (e.g., communications equipment).

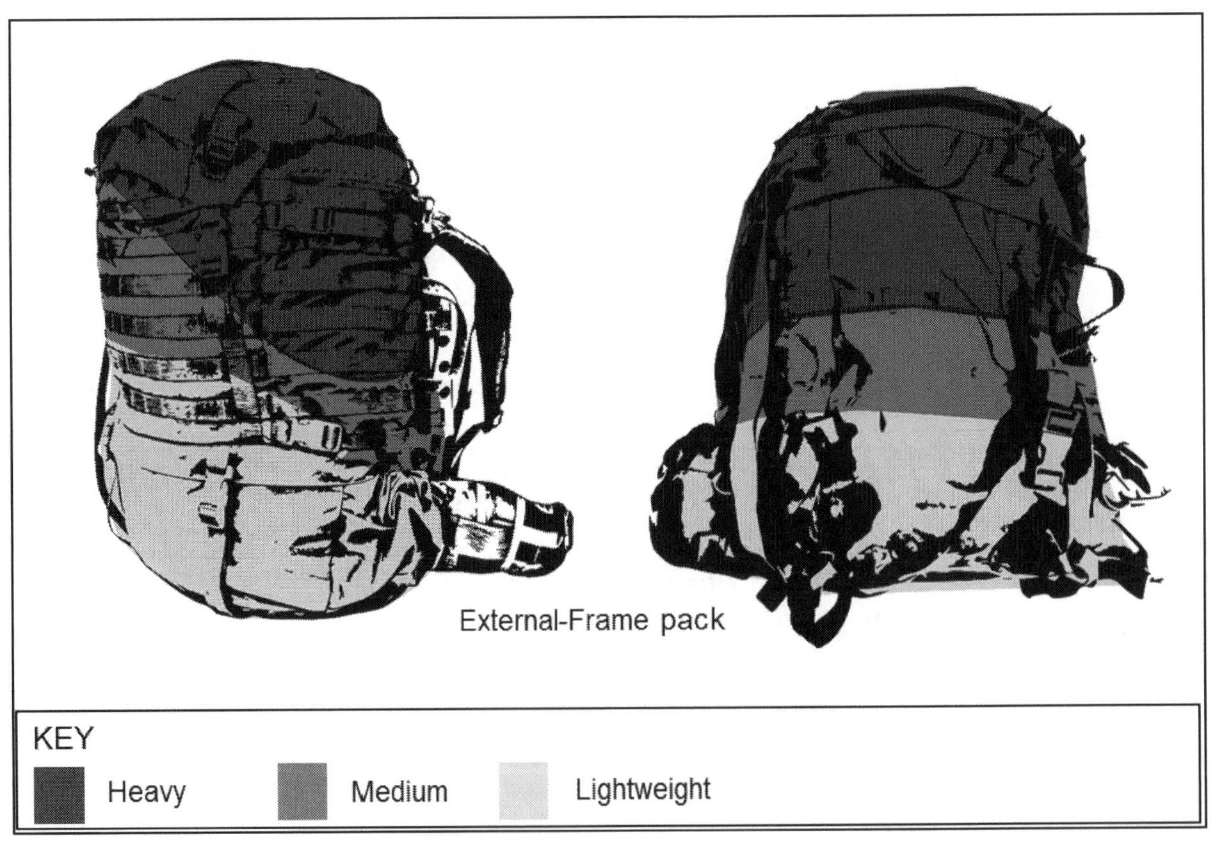

Figure F-2. Pack Sections and Distribution of Weight.

Top and Periphery

Heavy items should be in the top one-third so they are close to the spine in order to create a comfortable center of gravity. Softer, lower-weight but commonly used items (e.g., poncho liner, rain gear, and Goretex suit) can be wrapped around heavier items to prevent them from shifting. Items such as a flashlight, batteries, tarpaulin (i.e., poncho), navigation equipment, extra socks, camouflage paint, navigation equipment, and other smaller frequently used items should be packed in the zippered top pouch for easy and rapid access.

Exterior

Equipment attached to the exterior of the pack should be kept to a minimum to prevent creating instability in the load. External items can become snagged on brush and branches and can cause difficulty in moving through areas of dense vegetation. Externally packed equipment should be limited to the sleeping mat, entrenching tool, and sustenance and hydration pouches. Externally mounted equipment should be "dummy corded" to reduce the possibility of losing it.

Other Considerations

All empty spaces should be filled. Marines should maximize the use of available space to reduce the overall size of the pack. Once the pack is loaded, the compression straps on its exterior should be cinched tight to help pull the load closer to the spine and help keep the load from shifting.

SILENCING

Marines often operate in environments that are heavily occupied and controlled by the enemy. The rifle squad's aim is to meet the enemy on its terms, and thus it must take measures to avoid detection while traveling to and from the objective. This is done not only through the use of concealment and camouflage, but by silencing any equipment that may alert the enemy of the squad's presence during movement. Methods of silencing include—

- Packing gear in a way that prevents it from shifting inside of the pack.
- Wrapping exposed metal and hard plastic buckles in duct or electrical tape to reduce noise created by them striking other surfaces.
- Keeping canteens full and avoiding blowing air into hydration bladders to avoid the sloshing of water in a partially full container during movement. If a Marine takes a drink from a canteen, it should be passed around until it is empty.
- Restricting the wear of wet weather gear (i.e., Goretex) during movement to prevent the "swishing" sound it produces.

REPAIR

Marines live a rugged existence. Over time, operating in rough terrain can cause damage to their gear. Depending on the situation, surveying or exchange gear may not be possible for a prolonged period of time. Marines need to be prepared to perform field expedient repairs to any damaged gear in the field.

Utilities, helmet covers, packs, and any other gear exposed to the terrain for extended periods of time will regularly come in contact with branches, rocks, and other obstacles that can potentially cause it to snag and tear. Left untended, continued exposure to such obstacles will result in further damage. Marines should have a basic repair kit containing at the minimum—

- A sewing kit (i.e., needles, thread, patches, safety pins, and buttons).
- Duct tape.
- Adhesive glue (e.g., rubber cement).
- A bicycle tire repair kit.
- Zip ties.
- 550 cord.

These items are just the minimum required for performing basic repairs in the field. A Marine must be familiar with basic knots and sewing techniques to ensure that repairs are able to sustain continued use in the field. In the case of holes in wet weather gear, repairs can be made by placing a patch of duct tape on either side of the hole or tear once the piece of gear has been allowed to dry. Patching kits, such as those used on inner tubes, are an excellent means of repairing wet weather gear, but are seldom available to the Marine in the field. Marines may consider adding such kits to their packing list prior to departing on a mission.

Due to the frame being external to the issued pack, it can be expected to come into contact with logs, rocks, and other obstacles that can potentially cause it to snap and break. Not only does this prevent the pack from sitting properly on the Marine's back, but the frame can be expected to click as the two broken ends continuously slide past each other during movement, creating an unnatural sound that has the potential to be heard by enemy forces. The Marine can temporarily repair such breaks by finding a firm straight object, such as a branch or tent stake, and taping it to create a splint over the broken portion of the frame until the pack can be replaced.

FIELD STRIPPING

Field stripping is a technique used to reduce the amount of weight and space taken up by supplies. It is not uncommon for Marines to be issued numerous supplies. Original packaging takes up precious space and adds unnecessary weight to the Marine's load. Any unnecessary packing materials should be removed and discarded prior to stepping off on a mission.

Rations

Meals, ready to eat (MREs) are packaged to be able to survive being dropped out of an aircraft during resupply and to maintain an extended shelf life. To the Marine on the ground, this packaging is excessive in both weight and volume. A properly field stripped ration can take up as little as one-third the space of a factory packaged MRE. Upon drawing their rations, Marines should open each meal and discard the components they will not require. Remove and discard cardboard containers, as well as accessories and heaters the Marine opts not to use. Spoons can be broken to reduce size, or discarded entirely if the Marine prefers to squeeze out the ration when consuming it. When field stripping an MRE, it is important to keep in mind that the menus are designed to meet certain nutritional requirements and the rations themselves should not be discarded. Field stripped meals can be rolled into the individual meal pouch and taped shut. This tape provides an easy way to pack out the trash and prevent spills when the Marine has consumed the rations inside.

Batteries

Lithium and AA batteries are essential to the Marine's ability to conduct night operations and are often issued in cardboard boxes or stiff plastic containers. Remove batteries from packaging and place them in rows along the sticky side of a strip of duct tape, which is wrapped over the top of the row to form a tightly packed, thin bandoleer of batteries. When batteries are needed, the tape is either peeled back or the required number of batteries cut off of the stick. This minimizes the space used for storage while preventing loose batteries from getting lost among other gear.

BIVOUACKING

Selecting a Bivouac Site

Leaders should never pass up an opportunity to rest their Marines, as there is little way for them to anticipate when the next opportunity to do so will present itself. When selecting a site, certain important considerations must be made:

- The site should only be as large as the tactical situation permits.
- It should be located in a secluded area.
- It should take advantage of available natural camouflage and concealment.
- It should blend in with the surroundings.
- It should be irregularly shaped and have a low silhouette.
- It should be situated in an area that allows for escape routes with observable approaches; the squad should not be backed into a corner.
- It should be located in a position that will allow drainage in case of flash flooding.

Types of Shelters

In many environments, Marines may be able to sleep comfortably on the ground with nothing more than their sleeping mats and sleeping systems to shield them from the elements. However, in certain extreme climates or conditions, the sleeping systems alone may prove insufficient, and Marines may need to construct shelters to keep both themselves and their gear warm and dry. Shelters can be expeditiously established utilizing natural terrain and vegetation. If the time and situation permit, Marines can construct a shelter utilizing their tarpaulins (i.e., ponchos). The issued tarpaulin provides a lightweight and easily transported means to construct a shelter quickly and with minimal effort. When utilizing the tarpaulin, the camouflage patterned side is oriented outwards. As the water-resistant materials used in the tarpaulin's construction can produce a shine, it is important that Marines not rely on it alone to camouflage their positions. Lightweight natural materials that do not cause the shelter to dip or bow should be used, and natural overhead concealment should be sought out.

One-Person Shelters.

Flat-Top. A flat-top shelter is the preferred technique when the purpose of construction is to escape from the sun, or from excessive dew or frost from condensation that forms due to changing temperatures between day and night. The tarpaulin should be laid out flat over an area of terrain on which Marines intend to establish their bivouac. Parachute cord is tied to the grommets on the corners of the tarpaulin, and then tied to anchors (e.g., trees, brush, stakes) so that the tarpaulin is raised off the ground. To maximize protection from condensation and to aid in preventing detection, the tarpaulin should be close to the ground (between 12 to 18 inches), high enough for the Marine to store their gear and enter/exit by crawling.

Warning

In high temperate areas, the shelter should be constructed in two layers to avoid creating an oven effect that causes overheating. A second tarpaulin is laid over the first with at least 12 inches of space between the two, creating an "air cavity" for heat to accumulate and dissipate out the sides (see figure F-3).

When selecting a site for a flat-top shelter, the covered area should be relatively flat. Due to the larger area it covers, care should be taken to avoid detection by air through the use of natural and artificial camouflage.

Lean-To. When the purpose of construction is to provide shelter from rain or snow, the tarpaulin can be formed into a lean-to. The tarpaulin is laid out over the area the Marines intend to establish their bivouac, and parachute cord is tied to the grommets. The opposite ends of the cords are then

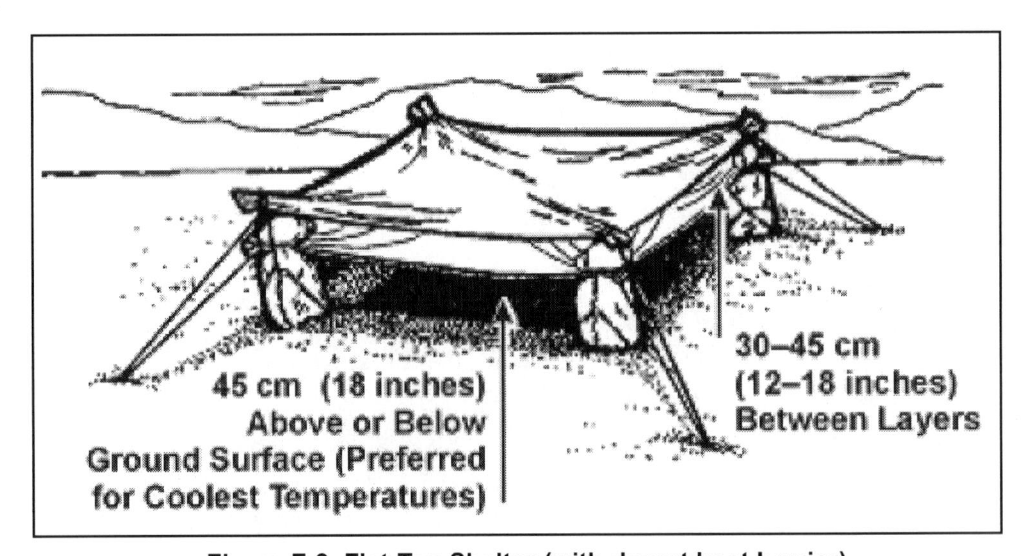

Figure F-3. Flat-Top Shelter (with desert heat barrier).

tied to trees or stakes, with the side of the lean-to facing the wind anchored at ground level, and the side facing away from the wind tied higher (see figure F-4), so that the tarpaulin lays at an angle that permits rain and snow to run off (see figure F-5). It is important that the tarpaulin be pulled tight when tied to prevent water from pooling in pockets of slack. The end on the ground can be kept in place by pinning it down with rocks or logs.

When selecting a site for a lean-to shelter, if the terrain permits, it should be established on the side of a hill to allow water to run off and prevent pooling. The Marine should sleep with their head toward the highest elevation (i.e., uphill).

Two-Person Shelter. For better coverage and protection from the elements, Marines may opt to establish two-person shelters by combining tarpaulins (i.e., ponchos). The one-person flat-top and lean-to methods can simply be extended, or the Marines may opt to construct an A-frame. When

Figure F-4. Lean-To Shelter (front view).

Figure F-5. Lean-To Shelter (side view).

constructing an A-frame, a pole, branch, or extension of rope or parachute cord is laid out at the center of the intended site to serve as a support. The tarpaulins are then laid out over the pole or cord with one partially overlapping the other. They should be overlapped at least two snap-lengths to better prevent water from penetrating the shelter. The overlapping snaps are then joined together and the Marines lift the center supporting material to the desired height and tie it tightly to trees, stakes, or other supporting materials. Parachute cord is tied to the grommets and the sides of the tarpaulins are pulled out and anchored down, forming a rectangular shelter (see figure F-6). They can be further held in place by pinning the bottom edges down with rocks or logs. To further waterproof the shelter, parachute cord can be tied to the grommets on the central overlapping ends, ran down along the slope of the sides, and staked down. If available, duct or other tape can be used to form a seal.

Field Expedient. If the time or situation do not permit construction of a proper shelter, natural terrain may be utilized and augmented as necessary. Thick overhead vegetation, caves, leeward (i.e., opposite the wind) hillsides, and other natural objects afford some level of protection from the elements and can be quickly reinforced with branches or covered with tarpaulins to provide a quick and hasty shelter from the elements. When reinforcing a site through such means, care must be taken to avoid oversaturating it in such a way that causes the site to stand out.

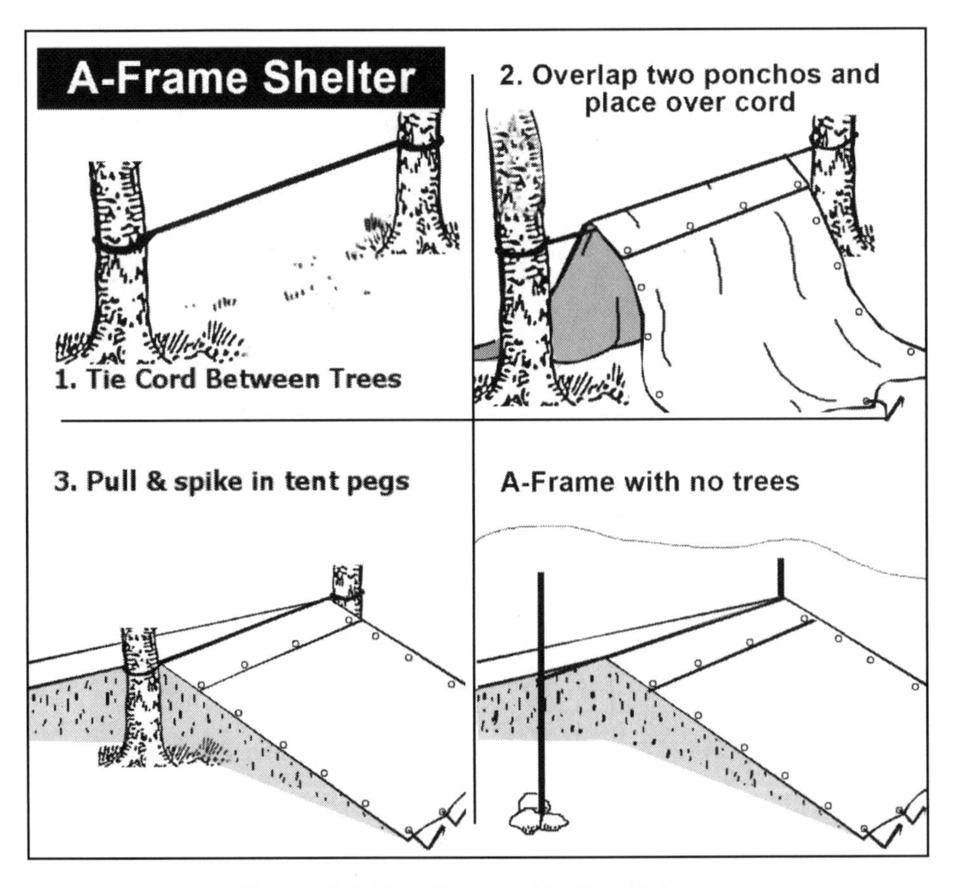

Figure F-6. Two-Person Shelter (A-frame).

FIELD HYGIENE

Due to the nature of its mission, the Marine rifle squad often operates for extended periods of time in dirty and adverse conditions, going without running water or soap for days or weeks at a time. Depending where in the world they are operating, they are often exposed to burning trash, foreign bacteria, and contaminated water on a daily basis. In order to remain effective, riflemen must maximize the use of available resources to remain hygienic and healthy.

Foot Care

Aside from their rifles, Marines' most important asset are their feet. Continuous daily movement and/or patrolling for extended periods of time and distances is not uncommon. At the end of the movement, the rifle squad is often expected to close with and destroy the enemy by assaulting through an objective, and, if needed, to conduct follow-on operations once the objective has been secured. Proper foot care is essential for the successful accomplishment of assigned missions.

Boots

Marines' boots are as essential in protecting their feet as their helmets are in protecting their heads. Authorized boots meet a set of specific standards intended both to stand up to the rigors of combat as well as to provide a certain degree of protection for the Marines' feet. When covering great distances, it is essential that Marines have well-fitting and serviceable boots.

Wear. When boots become overly worn, their ability to protect the wearer is diminished and the chances of foot and ankle injury increases. It is important that leaders inspect their Marines' boots regularly for signs of excessive wear. Boots determined to be unserviceable must be surveyed or replaced through the proper supply channels. Inspections should focus on checking that—

- Stitching is present and intact.
- Leather is supple and free of cracks and tears.
- The sole tread is not excessively worn down or coming unglued.
- Eyelets are not missing.
- The laces are serviceable.

Fit. Marines must ensure that their boots are properly fitted. Improper sizes can cause a Marine to develop blisters. When worn, laces should be cinched tightly to provide support to the ankles and prevent injuries if the Marine rolls an ankle on rough terrain. In the case of an ankle roll, tightly cinched boots help to inhibit swelling and allow the Marine to continue forward and keep pace with the unit.

Break-In. Boots should be properly broken in prior to use in a field environment to reduce the risk of injury. This can be done by initially wearing them during the conduct of low intensity activity or training, preferably when within friendly lines and prior to departing on a mission. Over time, intensity can be increased as the leather becomes supple and the boots begin to mold to the individual Marine's feet. In situations where time is not available, break-in can be expedited by immersing the boots in water and wearing them during the course of regular duties as they dry. As they dry, the leather will shrink slightly, molding to the shape of the Marine's feet.

Socks

Proper socks play as essential a role in Marines' foot care as their boots. Improperly fitting and overly-worn socks can cause the Marines' boots to rub excessively on the foot, causing blisters. In low temperatures, wet socks increase the risk of a cold-related injury such as frostbite and hypothermia. Leaders must strive to ensure that Marines should strive to maintain clean and dry socks, washing them by hand and air drying them if no other means to clean them are available. Torn or worn-out socks should be repaired with needle and thread until such a time as they can be replaced through the proper supply channels.

Basic Foot Care

Care of the foot itself is as important in the individual Marines' hygiene as proper boots and socks are. Feet should be kept clean and dry to prevent blisters and reduce the risk of cold-related injuries. Marines should never pass on an opportunity to dry their feet and change wet or damp socks. Leaders should inspect Marines' feet regularly for developing blisters.

A blister is a defense mechanism of the body. When the epidermis layer of the skin separates from the dermis, a pool of fluid collects between these layers while the skin re-grows from underneath. Blisters can be caused by chemical or physical injury. An example of a chemical injury would be an allergic reaction. Physical injury can be caused by the following:

- Improperly conditioned feet.
- Heat and moisture.
- Improperly fitting boots and/or socks.
- Friction and pressure.

Some of the signs and symptoms that blisters may be forming include—

- Fluid collection under the skin.
- Mild edema and erythema around the site.
- Sloughing of tissue, exposing subdermal tissue layer.
- Localized discomfort and/or pain.

Treating Blisters. Blisters need to be treated as soon as possible to mitigate any possible long-term effects on Marines' ability to accomplish the mission. Care of blisters should be handled by the unit's medical personnel whenever possible. Self-care should be used as a last resort to avoid possible infection. Blisters can be small, closed large blisters, or open blisters; each has its own treatment.

Small blisters usually need little to no treatment. Steps include—

- Clean the area with soap and water.
- Monitor the blister for signs and symptoms of infection.
- Apply a protective barrier (e.g. moleskin bandage) around the blister to prevent further irritation.

Closed, large blisters (if affecting Marines' mobility) should be treated as following:

- Wash the area around the blister with Betadine solution or an alcohol pad.
- Drain the blister as close to its edge as possible, then apply gentle pressure to the blister dome, expelling the clear fluid.
- Apply moleskin (in a donut shape) or cloth tape to the skin surrounding the blister, using tincture of benzoin as an adhesive.
- Dust the entire foot with foot powder to lessen friction and prevent adhesive from adhering to the socks.
- Monitor the blister for signs and symptoms of infection.

CAUTION
Do not put any adhesive directly on the blister.

Open blisters should be treated as follows:

- Wash the area with Betadine solution or clean it with soap and water.
- Remove any loose skin with a surgical blade or scissors.
- Apply moleskin (in a donut shape) or cloth tape to cover the skin surrounding the blister, using tincture of benzoin as an adhesive.
- Place a small amount of antibiotic ointment over the wound.
- Cut a Telfa pad and place it inside the moleskin or tape.
- Apply moleskin over the entire treated area, to include the surrounding skin.
- Monitor the blister for signs and symptoms of infection.

Body Care

Washing/Personal Hygiene

Regular washing is essential for good health. Six to seven gallons of water per day per Marine is considered the minimum consumption for all purposes. However, since it is often not possible to issue this amount when in the field, very strict individual and collective water discipline may be necessary. Marines should wash whenever the opportunity arises.

Cleanliness helps to protect the body against disease; however, cleanliness alone is not enough. By failing to take precautions against disease, or by drinking non-sterilized water, Marines who are thoroughly clean may yet endanger their health and life. Strict compliance with all rules and instructions for the protection of health is the only way to avoid diseases. Care must be given to the following areas.

Skin. A daily bathing is ideal, but not always possible. The opportunity to bathe should never be neglected. When possible, all parts of the body should be washed daily, paying particular attention to the parts where sweat collects (i.e., the armpits, around the waist, crotch, feet, and parts exposed to dirt), the face, and the hands. Hair should be kept short and clean.

Mouth and Teeth. Teeth should be cleaned at least once a day using a brush and toothpaste. In the absence of a tooth brush, Marines can use their fingers or a twig.

Ears. Ear trouble can be very painful and is particularly common in hot and humid climates. It can be contracted from swimming in dirty water, insects, and blowing sand; however, the collection of sweat and dirt in the ears is the most frequent cause. Ears should never be cleaned with such things as matches, twigs, or other articles that are likely to cause damage. Ear trouble needs medical attention immediately.

Hands and Feet. Hands continually come into contact with dirty and infected materials that may be transferred to food or other parts of the body. In particular, diseases and germs collect under the nails and can cause infections of the skin through scratching. Sweat and dirt collect on the feet, which render them particularly susceptible to infection through blisters and abrasions caused by ill-fitting boots and shrunken, badly worn, or mended socks. Points to take note of include—

- Keep fingernails short and clean.
- Wash hands before and after eating and using the toilet.
- Wash, dry, and—if possible—powder feet daily.
- Change socks daily and keep them well mended.
- Conduct foot inspections on a regular (if not a daily) basis.

Clothing and Footwear

The cleanliness of clothing is as important as the cleanliness of the body. Dirty clothing contains germs, which—when rubbed on the skin—can enter the body through abrasions and the pores of sweat glands and cause boils and other skin infections. Clothes should be changed and washed as often as possible. Sleeping systems and their liners should be washed and aired regularly, and inspected for insects and infestation.

Fatigue

Fatigue is an ever-present factor affecting Marines' performance, regardless of their degree of fitness, morale, and training. Leaders must continually be aware of this problem, particularly in themselves. Although the effects of fatigue are lessened by physical fitness, the ability to perform mental tasks can be affected.

WATER PURIFICATION

Untreated water can contain pathogenic agents which may include protozoa, bacteria, viruses, and some larvae of higher-order parasites, such as liver flukes and roundworms. Chemical pollutants such as pesticides, heavy metals, and synthetic organics may be present. Other components may affect taste, odor, and general appeal, including turbidity from soil or clay, color from acids or microscopic algae, odors from certain type of bacteria (particularly actinomycetes, which produce an earthy flavor), and saltiness from brackish or sea water.

Pathogenic viruses may also be found in water. The larvae of flukes are particularly dangerous in areas frequented by sheep, deer, or cattle. If such microscopic larvae are ingested, they can form potentially life-threatening cysts in the brain or liver.

In general, more human activity in an area, the greater the potential for contamination from sewage, surface runoff, or industrial pollutants. Groundwater pollution may occur from human activity or might be naturally occurring. Water should be collected as far upstream as possible above all known or anticipated risks. Pollution poses the lowest risk of contamination and is best suited to portable treatment methods.

Marines typically have the ability to purify water through the following methods: boiling, chemicals (e.g., purification tablets, iodine, or chlorine), or by using portable water filters. Marines should not rely on only one method to render water safe to drink. A combination of these methods—boiling, filtering, and chemical disinfection—is recommended whenever possible.

Boiling

Heat kills disease-causing micro-organisms, with higher temperatures and/or duration of boiling required to kill some pathogens. Sterilizing water (i.e., killing all living contaminants) is not necessary to make water safe to drink; Marines only need to render harmless enteric (i.e., intestinal) pathogens. Boiling does not remove most pollutants. Marines should boil water for a period of at least three minutes at altitudes below 6,000 feet, adding one minute for every 1,000 feet above that.

Filtration

When Marines utilize their issued pump-actuated hand-held filters, they must be aware of the following:

- The filters must be changed on a regular basis to ensure effectiveness.
- The pump's filter must be of the correct size (usually 0.2 to 0.3 micrometers) to remove pathogens.
- Filtering must be incorporated with chemicals (e.g. iodine, bleach, or chlorine) to remove bacteria.

Chemical Disinfection

Chemical disinfection, chiefly using chlorine and iodine, results from the oxidation of cellular structures and enzymes. The primary factors that determine the killing effect of chemicals are the number of microorganisms in the water compared to the concentration and the exposure time. Secondary factors include the pathogen species, water temperature, and organic contaminants.

Warning

Many species are extremely resistant to chemicals, and field inactivation may not be practical with bleach and iodine.

Iodine. Iodine is commonly added to water as a solution (typically found in Marines' first aid kits) to purify it. The iodine kills many (but not all) of the most common pathogens present in natural fresh water sources. Carrying iodine for water purification is an imperfect but lightweight solution for those who need to field-purify drinking water. This requires 30 minutes in relatively clear, warm water, but is considerably longer if the water is muddy or cold.

Note: The prolonged use of iodine is not recommended, as it can have adverse effects over an extended period.

Bleach. The recommended dose of bleach is two drops of sodium hypochlorite solution (i.e. regular, unscented chlorine bleach) mixed per one quart of water. Once mixed, the water must be allowed to stand for 30 minutes. The amount of bleach should be doubled if the water is cloudy, colored, or very cold. Afterwards, the water should have a slight chlorine odor. If it does not, repeat the dosage and allow the water to stand for another 15 minutes before use.

Terrain Models

The terrain model is the standard tool for communicating a combat order. It is a scaled-down model of the battlespace that squad leaders use in conjunction with issuing their order, the "I" step in BAMCIS (i.e., the troop leading steps). An effective terrain model allows the squad members to visualize the battlespace they will be operating in, the enemy they are engaging, and the scheme of maneuver they will execute.

Wise squad leaders evaluate their terrain model by asking the following question: "If someone with basic military knowledge were to look at my terrain model, could that person generally understand the terrain, the enemy and the friendly scheme of maneuver without explanation?" Table F-1 contains a partial list of items that should displayed on a terrain model.

Table F-1. Terrain Model Display Items.

North-seeking arrow	Main supply routes
Grid lines	Built-up areas
Boundaries	Vegetation
LZs	Water features
Targets	Known trails
Tactical control measures	Relief features

Squad leaders should ensure that all of their fire team leaders maintain a terrain model kit. The purpose of fire teams maintaining their own kits is to make them self-reliant when they are detached. When the squad is employed as an aggregated unit, squad leaders may combine the fire teams' terrain model kits to issue the order. Fire team leaders may choose to augment their kits with additional materials, but at minimum they should contain the items listed in table F-2.

Table F-2. Terrain Model Kit Contents.

Cords	
Length (feet)	**Purpose**
10 White	Grid lines
10 Black	Roads and trails
10 Blue	Rivers, streams, and bodies of water
10 Yellow	Phase lines and boundaries
10 Green	Utility purposes
Laminated Graphics and Symbols (Scale appropriate)	
Friendly and threat fire team, squad, and platoon symbols	
Tactical Control Measures: • Assembly Area • Attack Position • Assault Position • Line of Departure • Objective	
Light/heavy mortars, machine guns, and assault weapons	
Support by fire position	
Fire team and squad defensive positions	
Checkpoints, LPs, and OPs	
10 laminated blank cards for miscellaneous graphics and symbols (indirect fire targets, no-fire areas, etc.)	

When constructed properly, terrain models enhance squad leaders' ability to communicate the details and complexities of a plan to their squads. With a three-dimensional picture of the terrain to be covered during the assigned mission, a squad leader can create a mental image of the planned scheme of maneuver for the squad. The most important factor for terrain model construction is detail. Squad leaders should ensure their terrain models are as detailed as possible.

The proper procedure for constructing a terrain model is as follows:

- Use a map to identify the area required; include the entire area that is identified in the order. Identify any prominent or defining features within the area, such as mountains, hills, or water features. Ensure that any relevant areas of interest and adjacent units are included, particularly if they could potentially affect the proposed scheme of maneuver.
- Orient the terrain model with reference to the direction of movement. Ensure that north on the terrain model is always really north.
- The size of the terrain model should be appropriate for the squad and the mission. Ensure that the terrain model is large enough for the entire squad and attachments (if any) to gather around.

> ***Note:*** It is recommended that subordinate leaders are placed at the base of the terrain model, oriented in the direction of movement, when issuing orders and discussing details.

- The terrain model should accurately represent the terrain to be encountered within the area to be covered. The Marines must be able to determine at a glance the type of terrain they will be negotiating during the operation.
- Ensure that the terrain is proportional; use some sort of scale to establish common proportion for elevation and slope. Bring out the terrain to facilitate comparison to the map.
- All markings should coordinate with the colors used on a map. For instance, blue yarn should be used for water features, and black yarn should be used for roads or trails.
- To avoid wasting time reading grid coordinates several times during the order and to eliminate any possible error during transcription, write the grid coordinates on laminated white cards and include them on the terrain model. Ensure the labels can be seen from as far away as possible; provide Marines as much information as possible.
- Ensure that all units, tactical control measures, and features (i.e. both natural and man-made) are properly identified with the appropriate operational terms and graphics. Lack of attention to detail could result in Marines lacking confidence in the squad leader's abilities.

> ***Note:*** At times, it may be feasible to use a smaller terrain model specifically for the objective if it provides further clarity in briefing the order.

APPENDIX G
TACTICAL TASKS

OVERVIEW

Mission and tasking statements consist of answers to the questions who, what, when, where, and why. The Marine Corps maintains a standardized list of tactical tasks that answer the what portion of these tasking statements; they can be found in MCDP 1-0. These tactical tasks are the defined actions, based on unit capabilities, that a commander can take to accomplish the mission. They may be specified, implied, or essential.

These tactical tasks have nuanced and distinct meanings. They can be characterized or grouped as either enemy-oriented, terrain-oriented, friendly-oriented, or population-oriented, as shown in table G-1. A few, annotated by asterisks (*), can have multiple uses or meanings. In special circumstances, commanders can modify tasks to meet METT-T requirements. However, they must clearly state that they are departing from a task's standard meaning. One way this can be done is by prefacing the modified task with the statement, "*What I mean by* [modified task] *is . . .*" Commanders are not limited to the tactical tasks listed in MCDP 1-0 in specifying desired subordinate actions. Both the commander and subordinates must have a common understanding of what the commander is trying to convey in terms of the what and why of an operation.

Table G-1 lists tactical tasks per MCDP 1-0. The tactical tasks that squad leaders are most likely to be charged with when employed as a unit (plus attachments, when applicable) are highlighted in bold italic font. The descriptions of these select tasks are located after the table. They are for guided discussion only and are not official definitions of the terms in most cases. For the definitions, as applicable, see the *Department of Defense Dictionary of Military and Associated Terms*, and MCRP 1-10.2, *Marine Corps Supplement to the DOD Dictionary of Military and Associated Terms*. They are also contained in the glossary.

ENEMY-ORIENTED TACTICAL TASKS

The following tactical tasks focus friendly efforts on generating effects against enemy forces.

Ambush

A surprise attack by fire from concealed positions on a moving or temporarily halted enemy.

> ***Note:*** An ambush is fundamentally a type of attack, enemy-oriented, and is planned and executed accordingly.

Table G-1. Tactical Tasks.

<table>
<tr><th>Enemy-Oriented Tactical Tasks</th><th>Terrain-Oriented Tactical Tasks</th><th>Friendly-Oriented Tactical Tasks</th></tr>
<tr><td rowspan="3">***ambush***
attack by fire
block
breach*
bypass
canalize
contain*
corrupt
deceive
defeat
degrade
deny
destroy
disrupt
exploit
feint
fix
influence*
interdict
isolate
neutralize
penetrate
reconnoiter*
support by fire
suppress</td><td>***breach****
clear
control*
cordon*
occupy*
reconnoiter*
retain
secure*
seize</td><td>***cover***
disengage
displace
exfiltrate
follow and assume
follow and support
guard
protect
screen</td></tr>
<tr><th colspan="2">Population-Oriented Tactical Tasks</th></tr>
<tr><td>***advise***
assess the population
assist
build/restore infrastructure
contain*
control*
coordinate with civil authorities
cordon*</td><td>enable civil authorities
exclude
influence*
occupy*
reconnoiter*
secure*
train
transition to civil control</td></tr>
<tr><td colspan="3">KEY:
*Tactical tasks with multiple classifications and applications.
bold/italics: typically squad leader tasks</td></tr>
</table>

Attack by Fire

Fires (direct and indirect) in the physical domains and/or through the information environment to engage the enemy from a distance to destroy, fix, neutralize, or suppress.

> *Note:* Within physical domains, an attack by fire closely resembles the task of support by fire. The chief difference is that one unit conducts the support by fire task to support another unit as it maneuvers against the enemy.

Block

As a tactical task, to deny the enemy access to an area or prevent the enemy advance in a direction or along an avenue of approach. It may be for a specified time. As an obstacle effect, to integrate fire planning and obstacle effort to stop an attacker along a specific avenue of approach or to prevent an attacker from passing through an engagement area.

> *Note:* Block differs from the tactical task fix because a blocked enemy force can still move in another direction, it just cannot advance. A fixed enemy force cannot move.

Breach
To break through or secure a passage through an obstacle. See also terrain-oriented tactical tasks.

Deny
To hinder or prevent the enemy from using terrain, space, personnel, supplies, facilities, and/or specific capabilities.

Reconnoiter
To obtain, by visual observation or other methods, information about the activities and resources of an enemy or adversary. See also terrain- and population-oriented tactical tasks.

Support by Fire
Movement to a position where the maneuver force can engage the enemy by direct fire in support of another maneuvering force.

> *Note:* Support by fire closely resembles the task of attack by fire. The difference is a unit conducting attack by fire only uses direct and indirect fires. A unit conducting support by fire uses direct and indirect fires to support the maneuver of another friendly force.

Suppress
The transient or temporary degradation of an opposing force or the performance of a weapons system below the level needed to fulfill its mission objectives.

TERRAIN-ORIENTED TACTICAL TASKS

The following tactical tasks focus friendly efforts on achieving some sort of condition as it relates to terrain.

Breach
To break through or secure a passage through an obstacle. See also enemy-oriented tactical tasks.

Reconnoiter
To secure data, by visual observation or other methods, about the meteorological, hydrographic, or geographic characteristics of a particular area. See also enemy- and population-oriented tactical tasks.

Retain
To occupy and hold a terrain feature to ensure it is free of enemy occupation or use.

Secure
To gain possession of a position, terrain feature, piece of infrastructure, or civil asset, with or without force, and prevent its destruction or loss by enemy action. The attacking force may or may not have to physically occupy the area. See also population-oriented tactical tasks.

FRIENDLY FORCE-ORIENTED TACTICAL TASKS

The following tactical tasks focus efforts on supporting the actions of other friendly forces.

Cover
To conduct offensive and defensive actions independent of the main body to protect the covered force and develop the situation.

> *Note:* It is the tactical task associated with the security operation cover.

Disengage
To break contact with the enemy and move to a point where the enemy cannot observe nor engage friendly forces by direct fire.

Displace
To leave one position to take another while remaining in contact with the enemy.

> *Note:* Displace differs from the tactical task disengage in that units disengage to break contact with the enemy, while units displace to continue the mission or execute alternate missions.

Exfiltrate
To remove personnel or units from areas under enemy control by stealth.

Guard
To protect the main force by fighting to gain time while also observing and reporting information.

Screen
To observe, identify, and report information, and only fight in self-protection.

> *Note:* This is the tactical task associated with the security operation screen.

POPULATION-ORIENTED TACTICAL TASKS

The following tactical tasks focus friendly efforts on achieving some sort of condition as it relates to the population within the area of operations.

Advise
To improve the individual and unit capabilities and capacities of host nation security forces through the development of personal and professional relationships between United States and host nation forces.

Assess the Population
To evaluate the nature, situation, and attitudes of a designated population or elements of a population inhabiting the area of operations.

Coordinate with Civil Authorities
To interact with, maintain communication, and harmonize friendly military activities with those of other interorganizational agencies and coalition partners to achieve unity of effort.

Influence
To persuade the local population, including potential and known adversaries, within the operational area to support, cooperate with, or at least accept the friendly force presence, and to dissuade the population from interfering with operations. See also enemy-oriented tactical tasks.

Train
To teach designated skills or behaviors to improve the individual and unit capabilities and capacities of host nation security forces.

APPENDIX H
SQUAD UNIT TRAINING MANAGEMENT

Unit training management (i.e., UTM) is the Service's approach to ensuring units are adequately trained to accomplish their given missions. It is a task-focused method that relies on a hierarchy of mission essential tasks, their supporting subordinate tasks, and training standards. Squad leaders must be aware of their roles in the execution of their commanders' training plans in order to efficiently manage the time and resources allotted for training.

Commanders are responsible for developing and assigning the mission essential tasks that drive training. These are tied to the unit mission statements as well as tasks assigned to the unit in its associated operations and contingency plans. The commander's mission essential task list (METL) will encompass all the missions that unit is responsible to be ready to accomplish. Since the Marine Corps' approach to training focuses on wartime tasks, the METL will be composed of tasks the unit will be expected to conduct in combat. Supporting tasks, such as administrative or logistical tasks, are important, but they must ultimately serve the higher-order wartime task.

For further information beyond what is included in this appendix, refer to MCTP 8-10A, *Unit Training Management Guide*, and MCTP 8-10B, *How to Conduct Training*.

TRAINING PRINCIPALS

Train as You Fight

"Train as you fight" is the fundamental principal upon which all Marine Corps training is built. Combat is the ultimate test of training's effectiveness. Therefore, squad leaders must strive to train in a realistic manner, attempting to replicate the conditions under which they are expected to fight.

Leaders are responsible for training. Just as a battalion commander is responsible for the training of the command, squad leaders are responsible for the training of their squads.

Use Standards-Based Training

Training standards are published for each individual and unit type in the Service. Training based on common standards allows Marines to train, operate, maintain, and fight from a common perspective. These standards also form the foundational evaluation criteria by which Marines assess training effectiveness.

Use Performance-Oriented Training
Marines must be proficient in the basic skills required to perform their jobs under battlefield conditions. Individual training should be continual and is integrated into collective training. Marines are trained to meet published standards, and to maximize their designated training time.

Train as a Combined Arms Team
The strength of the MAGTF is in the use of combined arms teams. Rifle squads should seek to maximize the benefits of their organic combined arms assets throughout training. Additionally, they should seek to train alongside their common attachments and other enablers.

Train to Sustain Proficiency
It is not enough simply to meet a training standard and then move on to the next training task. Combat skills are perishable. An effective training plan continues to allow for the sustainment of the squad's skills as it progresses through various training events.

Train to Challenge
Marines build their battlefield confidence through executing challenging and meaningful training. Squad leaders must do what is best for their Marines, not what is comfortable for them. This further builds camaraderie and confidence.

Building a Training Plan

Training is broken down into collective and individual events. Collective events are those that require an entire unit. Individual events, when added together, enable the accomplishment of a collective task. The training events for the rifle squad are found in Navy Marine Corps Departmental Publication (NAVMC) 3500.44_, *Infantry Training and Readiness Manual.* This manual details which collective tasks serve the commander's mission essential task list. From those collective tasks, Marines can use the manual to identify which subordinate unit collective and individual training events are required to enable the commander's mission essential tasks.

A rifle company example is provided in the following paragraphs.

The battalion commander has stated that "Conduct Offensive Operations (MCT 1.6.1)" is one of the battalion's mission essential tasks. The company commanders are tasked to design and execute a training plan to support this mission essential task. Therefore, the company commanders will likely create a training plan that includes the company-level collective event, "Conduct a ground attack (INF-MAN-6001)" from the training and readiness (T&R) manual. The T&R manual shows that this event includes the chained event for the rifle platoon to "Conduct Ground Attack (INF-MAN-5001)." Likewise, this platoon event contains the chained event to "Conduct Ground Attack (INF-MAN-4001)" at the squad level (see figure H-1).

INF-MAN-4001: Conduct a ground attack (B, D)

SUPPORTED MET (S): 1,2

EVALUATION CODED: NO SUSTAINMENT INTERVAL: 3 months

CONDITION: Given a unit, attachment, an order, while motorized, mechanized or dismounted, and operating in the full range of environmental conditions, daylight and limited visibility.

STANDARD: To accomplish the mission and meet commander's intent.

Figure H-1. Sample From Infantry T&R Event: Conduct Ground Attack (INF-MAN-4001).

Once squad leaders have identified their event, they can then use the event components section to identify which individual tasks their units must train to in order to accomplish the collective training event. The squad leaders then move to the individual events section in the T&R manual to identify which events (organized by MOS) contribute to that higher-level event component.

Under the section for 0311 Riflemen (see figure H-2), the squad leader identifies the individual event "Perform individual actions in a fire team (0311-OFF-1001)." This individual event contains performance steps that can be found in a particular doctrinal publication, as identified in the references section of the event description. Events in the T&R manual also include the required facilities and resources needed to accomplish the training event, which can be used in planning and designing training.

This systems approach to training effectively layers key training events so that each event supports a number of warfighting tasks. Leaders can use this method to prioritize which events and components are most necessary to train to. Squad leaders must become familiar with this T&R event hierarchy and be able to identify and drill down on the relevant training components to support their commander's intent for training.

0322-OFF-1001: PERFORM INDIVIDUAL ACTIONS IN A FIRE TEAM (B,D)

EVALUATION CODED: NO SUSTAINMENT INTERVAL: 12 months

MOS PERFORMING: 0311

GRADES: PVT, PFC, LCPL, CPL

INITIAL TRAINING SETTING: FORMAL

CONDITION: Given an individual weapon, as a member of a fire team, while wearing a fighting load.

STANDARD: To arrive at the objective.

PERFORMANCE STEPS:

1. Perform individual actions as a member of a buddy team.
2. Perform individual actions during fire and movement.
3. Perform individual actions in a wedge.
4. Perform individual actions in a column.
5. Perform individual actions in skirmishers (right) (left).
6. Perform individual actions in an echelon (right) (left).
7. Perform individual actions as a part of the fighter-leader concept.
8. Perform individual actions as part of a base unit.

PREREQUISITE EVENTS: 0300-PAT-10

REFERENCES:

1. MCRP 3-10A.4 Marine Rifle Squad

SUPPORT REQUIREMENTS:
ORDNANCE:

DODIC		Quantity
A080	Cartridge, 5.56mm Blank x200 Single	20 rounds per weapon
L312	Signal, Illumination Ground White St	10
L594	Simulator, Projectile Ground Burst M	1 Simulator per Team

Figure H-2. Sample From Infantry T&R Manual: 0311 Rifleman Individual Event.

GLOSSARY

Section I: Abbreviations and Acronyms

AAV....amphibious assault vehicle

CAS....close air support
CASEVAC....casualty evacuation
CBRN....chemical, biological, radiological, and nuclear
CLIC....company level intelligence cell

FDC....fire direction center

GTL....gun-target line

HHQ....higher headquarters
HMMWV....high mobility multipurpose wheeled vehicle

IED....improvised explosive device
IO....information operations
IR....intelligence requirement

JFO....joint fires observer
JTAC....joint terminal attack controller

LP....listening post
LZ....landing zone

m....meter
mm....millimeter
MACO....marshalling area control officer
MAGTF....Marine air-ground task force
MCDP....Marine Corps doctrinal publication
MCRP....Marine Corps reference publication
MCTP....Marine Corps tactical publication
METT-T....mission, enemy, terrain and weather, troops and support available—time available
MRE....meal, ready to eat

NATO....North Atlantic Treaty Organization
NVD....night vision device

OP....observation post
ORP....objective rally point

*PCP...personnel checkpoint
PLD...probable line of deployment
PZ...pickup zone

QRF...quick reaction force

S-2... intelligence officer/office
SEAD...suppression of enemy air defenses
SOP...standing operating procedure

T&R...training and readiness
TRAP...tactical recovery of aircraft and personnel
TOT...time on target

UAS...unmanned aircraft system
US...United States

*VCP...vehicle checkpoint

Note: Abbreviations marked with an asterisk are valid for this publication only.

Section II: Terms and Definitions

ambush—A surprise attack by fire from concealed positions on a moving or temporarily halted enemy. (USMC Dictionary)

area reconnaissance—A directed effort to obtain detailed information concerning the terrain or enemy activity within a prescribed area such as a town, ridge line, woods, or other features critical to operations. (USMC Dictionary)

assault position—That position between the line of departure and the objective in an attack from which forces assault the objective. Ideally, it is the last covered and concealed position before reaching the objective (primarily used by dismounted infantry). (USMC Dictionary)

assembly area—An area in which a command is assembled preparatory to further action. (USMC Dictionary, part 1 of a 2-part definition)

assess the population—To evaluate the situation and attitudes of the civil population inhabiting the area of operations. This will likely be an ongoing task that friendly forces use to determine how and to what extent its own or enemy actions or environmental events are likely to affect the actions of the population. (USMC Dictionary)

attack by fire—Fires (direct and indirect) in the physical domains and/or through the information environment to engage the enemy from a distance to destroy, fix, neutralize, or suppress. (USMC Dictionary)

attack position—The last position occupied by the assault echelon before crossing the line of departure. (DOD Dictionary)

block—To deny the enemy access to an area or prevent enemy advance in a direction or along an avenue of approach. It may be for a specified time. (USMC Dictionary, part 1 of a 2-part definition)

breach—To break through or secure a passage through an obstacle. (USMC Dictionary)

combat patrol—A tactical unit that is sent out from the main body to engage in independent fighting. It may be to provide security or to harass, destroy, or capture enemy troops, equipment, or installations. Operations include raids, ambushes, and security missions. (USMC Dictionary)

contact patrols—Those combat patrols that establish and/or maintain contact to the front, flanks, or rear by contacting friendly forces at designated points; establishing contact with a friendly or enemy force when the definite location of the force is unknown; and maintaining contact with friendly or enemy forces. (USMC Dictionary)

coordinate with civil authorities—To harmonize military activities with those of other (nonmilitary) government agencies, nongovernmental organizations, and national or local host nation government entities in order to achieve unity of effort and facilitate meeting objectives. (USMC Dictionary)

deny—(See DOD Dictionary, denial measure, for core definition. Marine Corps amplification follows.) To hinder or prevent the enemy from using terrain, space, personnel, supplies, and/or facilities. (USMC Dictionary)

disengage—To break contact with the enemy and move to a point where the enemy cannot observe nor engage the unit by direct fire. (USMC Dictionary)

displace—To leave one position and take another. Forces may be displaced laterally to concentrate combat power in threatened areas. (USMC Dictionary)

final coordination line—A line used to coordinate the ceasing and shifting of supporting fires and the final deployment of the assault echelon in preparation for launching an assault against an enemy position. (USMC Dictionary)

guard—To protect the main force by fighting to gain time while also observing and reporting information. (USMC Dictionary, part 1 of a 2-part definition)

infiltration—The movement through or into an area or territory occupied by either friendly or enemy troops or organizations. The movement is made, either by small groups or by individuals at extended or irregular intervals. When used in connection with the enemy, it implies that contact is avoided. (USMC Dictionary, part 1 of a 2-part definition)

influence the population—To persuade the civil population within the area of operations to support or cooperate with operations by friendly forces, or, at a minimum, to accept the friendly force presence in the area of operations.

meeting engagement—A combat action that occurs when a moving force, incompletely deployed for battle, engages an enemy at an unexpected time and place. (USMC Dictionary)

probable line of deployment—An easily recognized line selected on the ground where attacking units deploy in line formation prior to beginning a night attack. Also called PLD. (USMC Dictionary)

reconnoiter—To obtain, by visual observation or other methods, information about the activities and resources of an enemy or potential enemy. (USMC Dictionary, part 1 of a 2-part definition)

retain—To occupy and hold a terrain feature to ensure it is free of enemy occupation or use. (USMC Dictionary)

route reconnaissance—A directed effort to obtain detailed information of a specified route and all terrain from which the enemy could influence movement along that route. (USMC Dictionary)

screen—A security element whose primary task is to observe, identify, and report information, and only fight in self-protection. (USMC Dictionary, part 1 of a 2-part definition)

secure—To gain possession of a position or terrain feature, with or without force, and to prevent its destruction or loss by enemy action. The attacking force may or may not have to physically occupy the area. (USMC Dictionary)

support by fire—To engage the enemy by direct fire to support a maneuvering force using overwatch or by establishing a base of fire. The supporting force does not capture enemy forces or terrain. (USMC Dictionary)

suppressive fire—Fires on or about a weapons system to degrade its performance below the level needed to fulfill its mission objectives, during the conduct of the fire mission. (USMC Dictionary)

zone reconnaissance—A directed effort to obtain detailed information on all routes, obstacles (to include chemical or radiological contamination), terrain, and enemy forces within a zone defined by boundaries. A zone reconnaissance normally is assigned when the enemy situation is vague or when information concerning cross-country trafficability is desired. (USMC Dictionary)

References

Joint Issuances

Joint Publications (JPs)
Department of Defense Dictionary of Military and Associated Terms

Miscellaneous
DOD Dictionary of Military and Associated Terms

Navy Publications

Navy Marine Corps Departmental Publications (NAVMCs)
3500.44_Infantry Training and Readiness Manual

Army Publications

Department of the Army Pamphlet
385-63 Range Safety

Marine Corps Publications

Marine Corps Doctrinal Publications (MCDPs)
1 Warfighting

1-0 Marine Corps Operations

Marine Corps Warfighting Publications (MCWPs)
3-03 Stability Operations

3-32 Marine Air-Ground Task Force Information Operations

Marine Corps Tactical Publications (MCTPs)
3-01A Scouting and Patrolling

3-01B Air Assault Operations

3-01C Machine Guns and Machine Gunnery

3-01E Sniping

3-10B Marine Corps Tank Employment

3-10C Employment of Amphibious Assault Vehicles

8-10A Unit Training Management Guide

8-10B How to Conduct Training

10-10A Multi-Service Tactics, Techniques, and Procedures for the Tactical Employment of Nonlethal Weapons (NLW)

10-10E MAGTF Nuclear, Biological, and Chemical Defense Operations

Marine Corps Reference Publications (MCRPs)
1-10.2 Marine Corps Supplement to the DOD Dictionary of Military and Associated Terms (USMC Dictionary)

2-10A.6 Ground Reconnaissance Operations

2-10A.7 Reconnaissance Reports Guide

3-31.6 Multi-Service Tactics, Techniques, and Procedures for the Joint Application of Firepower (JFIRE)

12-10B.1 Military Operations on Urbanized Terrain (MOUT)

Marine Corps Order (MCO)
3570.1C Range Safety

Made in the USA
Columbia, SC
29 July 2024

e57b0899-932c-4450-8c32-6952c7c9ff4bR01